# THE VASCULAR FLORA OF OHIO

*Volume Two*

THE

*Dicotyledoneae* of Ohio

PART 3.  ASTERACEAE

*This work is a project of the Ohio Flora Committee*
*of The Ohio Academy of Science, sponsored by the Academy*
*and supported, in part, by grants from the Ohio Department of Natural Resources*
*and Bowling Green State University*

T. RICHARD FISHER

# THE
# *Dicotyledoneae* of Ohio

PART 3. ASTERACEAE

*Original drawings by Sharon Ames Glett*

OHIO STATE UNIVERSITY PRESS: Columbus

Endpaper Source: Miami University, Oxford, Ohio

Library of Congress Cataloging-in-Publication Data

The Dicotyledoneae of Ohio.
     p.  cm.—(The Vascular flora of Ohio; v. 2, pt. 3)
  "A project of the Ohio Flora Committee of The Ohio Academy of Science,
sponsored by the Academy"—Half t.p.
  Bibliography: p.
  Includes Index.
  Contents: —Pt. 3. Asteraceae / T. Richard Fisher; original drawings by
Sharon Ames Glett.
  ISBN 0-8142-0446-5 (pt. 3)
  1. Dicotyledons—Ohio—Identification. 2. Dicotyledons—Ohio—Geographical
distribution—Maps. 3. Botany—Ohio. 4. Phytogeography—Ohio—Maps.
I. Fisher, T. Richard, 1921–  II. Ohio Academy of Science. Ohio Flora
Committee. III. Series.
QK180.V35 vol. 2, pt. 3. etc.
581.9771 s—dc19
[583'.09771]                                     87–31257
                                                     CIP

Printed in the U.S.A.

# FOREWORD

Asters, bachelor's-buttons, chrysanthemums, dahlias, daisies, dandelions, goldenrods, ironweeds, marigolds, zinnias—the composites color our gardens and fields and the natural landscape with a full spectrum of vivid hues. Everyone knows at least one or two species. Artichokes, lettuce, and sunflowers provide items for the human diet. Ragweed causes an annual period of suffering for those allergic to its pollen. Canada thistle, another pernicious weed, invades pastures and cultivated fields, crowding out both grasses and crops. With some 20,000 species, the composites form one of the world's largest plant families. In this book, Dr. Fisher identifies and describes those members of the family that are a part of the Ohio flora.

The family has two scientific names: Asteraceae and Compositae. This anomaly is permitted by the rules of the *International Code of Botanical Nomenclature* because the usage of both names has become common in the scientific literature. The two are synonymous, and the use of either is equally correct.

The family is generally accepted by students of plant phylogeny as the most advanced, or highly evolved, of all dicotyledons. The plants' most remarkable structural feature is a tight inflorescence ("head") of flowers which simulates a single flower. What appears to be one blossom proves upon careful examination or dissection to be in fact a cluster of a few to a few hundred small flowers. Within the cluster a division of labor often occurs in which some flowers are adapted for attracting insect pollinators and others for producing either pollen or seeds, paralleling in function also the separate parts of a single flower.

The family is a highly successful one as measured by the number of species and its geographical distribution. Its species have moved into nearly every part of the earth inhabitable by flowering plants, from lakes to deserts, from prairies to woodlands, and from tropical to arctic and alpine climates. In line with this general situation, composites grow in every plant habitat in Ohio.

One-third of the Ohio composites are alien, mostly naturalized from Eurasia. Their frequency varies from a species such as *Heterotheca camporum*, golden aster, known from but a single site, to *Tussilago farfara*, coltsfoot, frequent in the eastern half of the state but scarce westward, to the ubiquitous *Taraxacum officinale*, dandelion. Alien species grow best in disturbed habitats, and composites are often the most conspicuous and showiest members of the weed communities inhabiting these sites. They are joined there by a few native composite species that also are able to compete successfully in disturbed places.

The other two-thirds of the species are native, or indigenous, to Ohio.

Some occur throughout the state. For example, *Solidago canadensis*, Canada goldenrod, and *Vernonia gigantea*, tall ironweed, which thrive in a great diversity of habitats, are reported from virtually every county. *Eupatorium perfoliatum*, common boneset, and *Verbesina alternifolia*, wing-stem, have a similar broad distribution because of the widespread occurrence of the moist, open fields and thickets they inhabit.

Many native species grow only in a particular part of the state. Their distribution maps presented here display an interesting set of patterns, often defining some particular aspect of Ohio's physiography.

*Eupatorium rotundifolium*, roundleaf thoroughwort, and *Coreopsis major*, a species of tickseed, are representatives of a group of species whose distribution is limited to the southernmost part of Ohio's unglaciated Allegheny Plateau. This region has overall floristic affinities with the Appalachians and is home to a number of species that are outliers of the Appalachian flora.

Other native composites are among the most attractive members of Ohio's prairie flora. Occurring chiefly in the western half of the state, the distribution records of these plants come from present and past remnants of the tall-grass prairie that once extended across the Mississippi River and eastward through the central parts of Illinois, Indiana, and Ohio. Among Ohio's showy prairie composites are the six species of *Liatris*, blazing-star, and *Echinacea purpurea*, purple coneflower, *Silphium terebinthinaceum*, prairie-dock, and *Ratibida pinnata*, prairie coneflower.

*Solidago gymnospermoides*, a species of goldenrod, and *Artemisia caudata*, a species of wormwood, are two of a small group of species whose Ohio distribution is limited to northern sites near Lake Erie. In the southern part of the state, *Antennaria solitaria*, single-headed everlasting, and *Rudbeckia fulgida*, orange coneflower, represent a group of species which reach in southern Ohio the northernmost limit of their range. Still other Ohio composites have an eastern or western distribution. *Prenanthes racemosa*, wild lettuce, grows in the calcareous soils found chiefly in western counties, while *Hieracium paniculatum*, panicled hawkweed, is part of a group of species found primarily in the acidic soils of eastern counties.

**Ohio Flora Project**

The Ohio Flora Project originated in 1950, under the leadership of the late Dr. E. Lucy Braun and the aegis of The Ohio Academy of Science. The project has served as a stimulus for a major amount of field work in Ohio, producing tens of thousands of new herbarium specimens from scores of individual research projects. The result is the acquisition of a detailed knowledge of the flora far greater than that known in 1950, and one that places Ohio's flora among the best known of any of the 50 states.

Dr. Braun prepared two volumes on the flora, *The Woody Plants of Ohio* and *The Monocotyledoneae*. The first was published by the Ohio State University Press in 1960, the second in 1967. Since then, more than fifty journal articles and theses on various aspects of the flora have resulted from the research projects noted above. Many of these are cited in my 1984 review, "Ohio's Herbaria and the Ohio Flora Project." Reference to that paper and other publications mentioned in this foreword are listed in the *Literature Cited* section at the end of this book.

Several major works on the flora have appeared since 1967. In 1971, Clara G. Weishaupt published the third edition of her widely used set of keys, *Vascular Plants of Ohio*. In 1977, *The Vascular Plants of Unglaciated Ohio*, by Allison W. Cusick and Gene M. Silberhorn, was published by the Ohio Biological Survey. A companion volume, *The Flora of the Glaciated Allegheny Plateau Region of Ohio*, by Barbara K. Andreas, is currently in press.

Other publications have resulted from the recent attention given Ohio's endangered plant species. The many new herbarium specimens collected since 1950 provided basic data on the flora itself and formed a sound data base for determinations of those Ohio plant species that were in danger of becoming extirpated. In 1982, a work that I edited, *Endangered and Threatened Plants of Ohio*, was published by the Ohio Biological Survey. It included individual sections written by William Adams, Marvin L. Roberts, Ray E. Showman, Jerry A. Snider, Ronald L. Stuckey, and myself. Together the sections treated all of the flora, both vascular and nonvascular plants, except for algae and fungi. In addition to identifying those species in jeopardy, the authors presented valuable lists of excluded species, ones that had been incorrectly attributed to the Ohio flora by past workers. In 1984, *Ohio Endangered and Threatened Vascular Plants*, edited by Robert M. McCance, Jr., and James F. Burns, was published by the Division of Natural Areas and Preserves of the Ohio Department of Natural Resources. It contained comprehensive abstracts for 357 species. That agency has also in recent alternate years (1982, 1984, 1986) issued updated status lists of extirpated, endangered, threatened, and potentially threatened species, each revision based on data resulting from new field work.

A long-term goal of the Ohio Flora Project has been the publication of three books on Ohio's dicotyledons (dicots) to complement Braun's volume on the monocotyledons (monocots), thereby completing a series covering all the flowering plants. Present plans are for a single dicot volume, to be composed of three separately issued parts. As was the case with the monocots, the families of dicots will be presented in the order used by M. L. Fernald in the 8th edition of *Gray's Manual of Botany*.

It is anticipated that the three parts of the dicot volume will be published in reverse order. *The Dicotyledoneae of Ohio. Part 1. Saururaceae through Leguminosae*, scheduled to appear last, is currently under preparation by Dr. John J. Furlow. This part will include introductory material on the Ohio dicots as a whole. I am now writing *Part 2. Linaceae through Campanulaceae*, scheduled to appear second. Publication of the dicot series begins with this book by Dr. T. Richard Fisher, *Part 3. Asteraceae*.

I thank Charles C. King, David B. MacLean, Richard E. Moseley, Jr., John H. Olive, David E. Todt, Ralph W. Wisniewski, and the late Carson Stauffer for valuable assistance each gave the Ohio Flora Project in recent years. I appreciate the cooperation of the members of the 113th (1979–80), 115th (1983–84), and 116th (1985–86) Ohio General Assemblies in providing financial support for the Project, even under tight budget conditions. Their support made this book and those that will follow possible.

Tom S. Cooperrider
Chairman, Ohio Flora Committee,
The Ohio Academy of Science, and
Department of Biological Sciences,
Kent State University

# PREFACE

This part of the *Vascular Flora of Ohio* was prepared using the guidelines established by the late Dr. E. Lucy Braun in the preparation of Volume 1, *The Monocotyledoneae* (Braun, 1967), and in her earlier work, *The Woody Plants of Ohio* (Braun, 1961), and adopted by the Ohio Flora Committee (see Cooperrider, 1984). Completeness, accuracy, and usability, not only by the professional botanist but also by the serious amateur, were kept uppermost in mind. The introductory section should be consulted, particularly the part dealing with general morphology, where I have discussed in detail the characteristics that are peculiar to the Asteraceae. These characteristics must be understood thoroughly, since they form the basis of the keys to the tribes, genera, and species.

I extend my sincere appreciation to my many friends in Ohio, especially to the curators of the herbaria for their hospitality and cooperation during my visits to their institutions. Without the consultation and examination of specimens, projects of this nature could not be completed. The illustrations were made from herbarium material and, as often as possible, from living plants, mostly by Sharon Ames Glett, to whom I am indebted not only for her outstanding artistic talent but also for her pleasant personality and high standards. I thank Ellen Powell for preparing the illustrations for the twenty-one species of *Solidago*. Thanks are extended also to Ann Otley who kindly typed portions of the manuscript and aided in the proofreading. My dear wife, Charlotte, provided me with encouragement and support during the several years I have worked on this project and even accompanied me on several collecting trips within Ohio.

I wish to express my gratitude to the Ohio Department of Natural Resources and to Bowling Green State University for their support by way of grants that provided me with a leave of absence in order to work on the project. I thank especially Richard E. Moseley, Jr., Chief of the ODNR's Division of Natural Areas and Preserves, and Dr. Charles C. King, Director of The Ohio Biological Survey, for their unfailing support of the Ohio Flora Project.

Finally, I extend my sincerest appreciation to the reviewers of the manuscript for their thoughtful comments and criticisms: Dr. Tom S. Cooperrider, Kent State University; Dr. John J. Furlow, The Ohio State University; Dr. James S. Pringle, The Royal Botanical Gardens, Hamilton, Ontario; Dr. Dale M. Smith, The University of California at Santa Barbara; and Dr. Ronald L. Stuckey, The Ohio State University; and to Dr. Cooperrider and the Ohio State University Press staff for editorial services.

Department of Biological Sciences, Bowling Green State University

# CONTENTS

# INTRODUCTION

This part of the *Vascular Flora of Ohio* covers the native and naturalized species of the Asteraceae (Compositae) that are known to occur in Ohio. Several local adventive species and a few garden escapes and ornamentals are also included.

There is a distribution map for each native and naturalized species. A dot in a county means that I have seen an herbarium specimen from that county. I have not included any sight records or any sight distribution reports that are not vouched by a specimen preserved and deposited in an herbarium. One specimen for each species from each county was selected and stamped "Annotated and mapped, Vol. 4, Flora of Ohio." In this manner a record was made that forms the basic documentation of the distribution for any particular species. As a matter of routine herbarium practice, all specimens examined were annotated by me with another stamp when I agreed with the existing identification. When I encountered an incorrect identification I corrected it, and I updated the nomenclature when a specimen was identified with an older name no longer in use.

Specimens were examined from the herbaria of The Ohio State University, Ohio University, Kent State University, Miami University, The University of Cincinnati, Bowling Green State University, and from several smaller collections. As specimens of the various herbaria were examined, species distribution records were made on 5-inch-x-7-inch cards indicating the county where the specimen was collected and the herbarium where the specimen is deposited. Upon the publication of this volume, the master file of these cards will be placed on deposit in the herbarium of The Ohio State University, Columbus, Ohio.

During the study for the preparation of this book, it became apparent that in some genera variation and intergradation of characteristics between species is not uncommon. Resolution of these situations requires biosystematic investigations beyond the scope of this study, and I have attempted to call attention to these problems in the comments following the treatment of species. Regardless of the cause or causes of this variation, the result is that species lines in these genera are sometimes obscured to such an extent that precise species identification is difficult. In the genus *Liatris*, for example, specimens approaching *L. scabra* may be found in various Ohio herbaria, but they appear to be hybrids with *L. aspera* or possibly some other species. As Braun (1967) pointed out, we are often dealing with "population-variation, hybridization and introgression . . . , and genetic races." I concur with this general observation. I believe that one of the objectives in the preparation of state floras is to call attention to variation when it is discovered. Determining the causes of such variation requires separate, detailed biosystematic studies. The naming of interspecific hybrids is controlled by rules of the *International Code of Botanical Nomenclature* (Voss, 1983). These are discussed in a special section under the genus *Helianthus*, p. 154.

The nomenclature and classification used here are, for the most part, those of *Gray's Manual of Botany*, 8th edition (Fernald, 1950), although in many cases I have adopted the taxonomy of Cronquist in *The New Britton and Brown Illustrated Flora of the Northeastern United States and Adjacent Canada* (Gleason, 1952). In still other instances, I have departed from both in order to follow more recent revisionary or monographic works. For example, the treatment of *Antennaria* is based on the excellent work of Bayer and Stebbins (1982). The treatment of several taxa is based on Cronquist's more recent work (1980) on the flora of southeastern United States. The studies of Steyermark (1963) and Weishaupt (1971) have also been helpful on certain matters.

Readers should consult Braun (1961, 1967) for her excellent summaries of the physiographic features of Ohio and their influence on plant distribution within the state.


## FAMILY DESCRIPTION

The family Asteraceae, often referred to as the Sunflower Family or the Composite Family and containing in excess of 20,000 species, is one of the largest flowering plant families in the world. The family reaches its maximum diversity in the temperate zone. In Ohio, 75 genera and 276 species are represented in the flora.

One of the most characteristic features of the family is the distinctive inflorescence, which consists of flowers borne in a dense head. These heads often superficially resemble a single flower and are thus often misunderstood by the layman and uninitiated botanist. The individual flowers often impart a pleasing color to the heads, and as a result many species are cultivated as ornamentals.

*General Morphology*

Because the family has a distinct type of inflorescence and is unique in several other respects (Figs. 1, 2), a slightly different descriptive terminology has developed about it. The most important of these terms are discussed below. The various features of the heads and flowers are described as they are most often found in Ohio. There are exceptions to many of the generalizations.

The species of Asteraceae, at least those in Ohio, most often either have perfect flowers (stamens and carpels in the same flower) or are monoecious (with separate staminate and carpellate flowers on the same plant), and are only rarely dioecious (having staminate and carpellate flowers on separate plants). In some cases the flowers are neutral (producing neither stamens nor carpels).

The small flowers are aggregated in a dense head on a common receptacle. The heads may either be solitary or grouped to form various types of inflorescences. As used in the keys and descriptions, the term inflorescence refers to a cluster of heads, and not to the individual heads themselves. The receptacles may be variously shaped. They are usually flattened, but are sometimes conical or columnar. The receptacle is surrounded by an involucre of one or more series of distinct or variously connate (fused) bracts, called phyllaries. The phyllaries also may be variously shaped, ranging from ones that are leaflike to ones prolonged into sharp spines, to others that are reduced in size and obsolete.

The ovary of the individual flower is always inferior, and consequently the flowers are epigynous. Usually each flower of the head is subtended by a bract (or pale), collectively called chaff, which may closely embrace the inferior ovary. The apex of the bract is sometimes brilliantly colored. When bracts are absent, the receptacle is described as naked.

Bristles, awns, teeth, or scales are often present at the summit of the ovary. These structures are referred to as the pappus, and they occupy the position of the calyx in flowers of other plant families. The pappus is quite variable (Fig. 3). It may be composed of simple bristles or of plumose bristles (those with fine secondary hairs), attached in one or more series, or it may consist of scales and stout awns, or be merely a projecting ridge, or have some combination of these features. Or the pappus may be entirely absent.

The corollas are of two basic types (in our flora). In the simplest, and probably the most primitive, the corolla is tubular, straight, actinomorphic, and usually has five short terminal lobes. The second type, which is usually more showy, is tubular only at the base, above which it flattens quickly into a zygomorphic, ligulate corolla. The distal portion of the ligulate corolla is usually at least partially divided into lobes or teeth.

A single head may be discoid, ligulate, or radiate. A head is discoid when it is composed only of tubular flowers, ligulate (Fig. 2) when it is composed only of ligulate flowers with flattened corollas, and radiate (Fig. 1) when it is composed of both tubular and ligulate flowers. When the head is radiate, the tubular flowers are located centrally and the ligulate flowers occupy a position around the margin of the head and are referred to as marginal flowers, or more often as ray flowers or rays. When ligulate flowers are present on the periphery of the head, they are often neutral (lacking stamens, ovaries, and styles) or ovulate only and hence seed-forming.

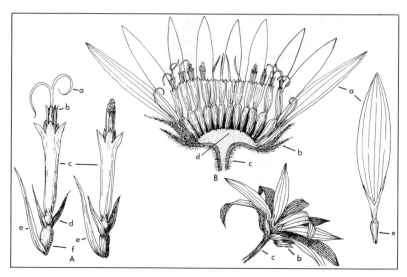

Figure 1. A. Tubular flowers: a) style branch and stigma, b) anthers, c) corolla, d) pappus, e) subtending bract or pale (chaff), f) ovary. B. Radiate head: a) ligulate corolla of marginal or ray flower, b) phyllary, c) peduncle, d) receptacle, e) ovary

In some genera, the tubular flowers produce only functional stamens, even though normal-appearing ovaries may be present. When this situation occurs, a parthenocarpic fruit may develop without a seed. These flowers should not be referred to as sexually sterile because they are a part of the sporophyte generation. Biologically a sporophyte is strictly nonsexual; only the gameto-phyte generation contains sexual structures. A different condition exists in the flowers of certain genera, *Heliopsis*, for example, in which the ligulate (ray) flowers are ovulate only. In this case, styles are present and are receptive to pollen and consequently a seed is formed in each fruit at maturity. In some other genera, those of the tribe Lactuceae, the head may be composed entirely of perfect, ligulate flowers.

The stamen number in each flower is five. The stamens are borne on the corolla tube (epipetalous) and are connate by their anthers, although often only weakly so. The bases of the anthers are various, usually caudate or sagit-tate or rounded, but relatively constant within tribes.

The style branches are usually two, reflecting the number of carpels in the gynoecium. As the style matures, usually after pollen maturation, elongation occurs causing the style branches to be pushed out through the corolla tube. On the way, they pass the ring of anthers positioned inside the corolla tube. Because the pollen sacs of the anthers open inwardly, the emerging style branches pick up large amounts of mature pollen and carry it upward and

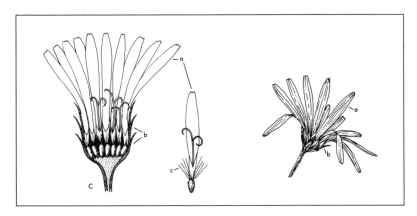

Figure 2. C. Ligulate head: a) ligulate corolla, b) phyllary, c) pappus

Figure 3. Pappus types: a) pappus absent, b) apex entire and coroniform, c) toothed crown, d) scales, e) awned scales, f) awns of 2 lengths, g) awns retrorsely barbed, h) simple bristles, i) plumose bristles, j) villous ring at base of bristles, k) achene winged, notched, l) achene beaked with awns above, m) pappus bristles in 2 series.

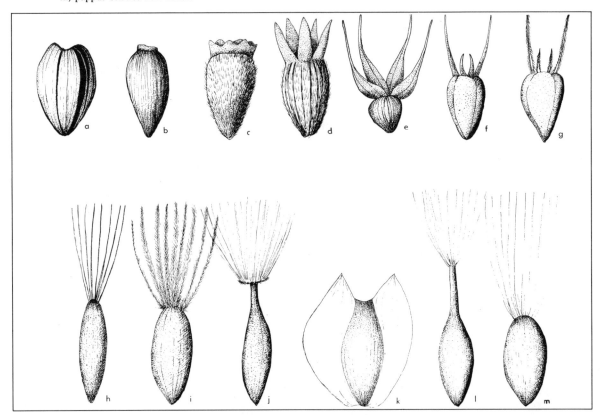

beyond the corolla tube from where it may be dispersed. The stigmatic area of the style is generally on the inner surface and is neither exposed to nor receptive to pollen until after it emerges above the corolla tube. The various types of style branches characteristic of the several tribes occur only in the tubular flowers, because those of the marginal or ray flowers are often reduced or undivided.

Although the gynoecium consists of two united carpels, the ovary is unilocular and bears a single ovule. The fruit is an achene, frequently compressed or obcompressed and often crowned or appendaged with any one of several persistent pappus types.

Most botanists are in agreement that this family is the most evolutionarily advanced of the dicotyledons.

# KEYS

Two different approaches may be taken to the identification of taxa included in this manual. In the first, a plant may be identified as a member of a particular tribe, see "Key to Tribes of Asteraceae" immediately below. Once the tribe to which the plant belongs has been determined, merely turn to the appropriate page in the text where a key to the genera of that tribe is provided.

The second approach to the identification of taxa is to use the section, "Key to Assemblages of Genera," p. 8. The genera of the family have been grouped into six artificial assemblages based on easily observed characters or combinations of characters. A key to genera is found under each of these six assemblages.

## KEY TO TRIBES OF ASTERACEAE

a.  Flowers all ligulate, perfect; plants with milky juice................
        .................TRIBE X. LACTUCEAE (CICHORIEAE), p. 224
aa. Flowers all tubular, or heads with both tubular and ligulate flowers; ray
    flowers imperfect (staminate or carpellate) or neutral; plants with or with-
    out milky juice.
  b.  Receptacles densely bristly (or honeycombed and slightly bristly); true
     ray flowers absent, but peripheral flowers sometimes enlarged; styles
     thickened and hairy below bifurcation ........................
     .............................TRIBE IX. CYNAREAE, p. 207.
  bb. Receptacles naked or chaffy, or if bristly, ray flowers present.
    c.  Involucral bracts green or only the tips or margins scarious; anthers
      truncate, rounded, or sagittate at base, but not caudate.
     d.  Receptacles naked or rarely bristly; pappus often with capillary
       bristles.
      e.  Involucral bracts imbricated in 2–many series.
      f.  Ray flowers absent; flowers all perfect, never bright yellow.
       g.  Style branches filiform; leaves alternate; pappus never plumose,
         nor heads in spicate to subracemose clusters.................
         .........................TRIBE I. VERNONIEAE, p. 16.
       gg. Style branches thick, obtuse; leaves opposite, subopposite, or
         whorled, or if alternate, either the pappus plumose or the clusters
         of heads spicate to subracemose .........................
         ........................TRIBE II. EUPATORIEAE, p. 21.
      ff.  Ray flowers usually present; if absent, the flowers bright yellow or the
       marginal carpellate only .........TRIBE III. ASTEREAE, p. 41.

ee. Involucral bracts only slightly if at all imbricated.

f. Pappus of scales or pappus absent . . . . . . . . . . . . . . . . . . . . . . . . . . . .

. . . . . . . . . . . . . . . . . . . . . . . . . . .TRIBE VI. HELENIEAE, p. 174.

ff. Pappus of capillary bristles . .TRIBE VIII. SENECIONEAE, p. 194.

dd. Receptacles chaffy, each flower in the axil of a bract; pappus without capillary bristles. . . . . . . . . . . . .TRIBE V. HELIANTHEAE, p. 112.

cc. Involucral bracts, or some of them, scarious, or the anthers caudate.

d. Pappus usually present, capillary; anthers caudate . . . . . . . . . . . . . . .

. . . . . . . . . . . . . . . . . . . . . . . . . . . . .TRIBE IV. INULEAE, p. 102.

dd. Pappus reduced or absent, not capillary; anthers not caudate . . . . . . .

. . . . . . . . . . . . . . . . . . . . . .TRIBE VII. ANTHEMIDEAE, p. 179.

# KEY TO
# ASSEMBLAGES OF GENERA

a. Heads discoid or radiate.

b. Heads discoid.

c. Pappus capillary. . . . . . . . . . . . . . . . . . . . . . . . . . . .Assemblage 1, p. 8.

cc. Pappus none, or, if present, not capillary . . . . . . .Assemblage 2, p. 10.

bb. Heads radiate.

c. Pappus capillary, or pappus absent or nearly so.

d. Pappus capillary. . . . . . . . . . . . . . . . . . . . . . . . .Assemblage 3, p. 11.

dd. Pappus absent, or a short crown, or of a few weak teeth, awns or scales that correspond to the angles of the achenes, sometimes with tiny bristles or scales between the angles . . . . . . . . .Assemblage 5, p. 12.

cc. Pappus a circle of awns, or chaffy or rigid bristles, or scales sometimes dissected into bristles, often deciduous . . . . . . . .Assemblage 4, p. 12.

aa. Heads ligulate . . . . . . . . . . . . . . . . . . . . . . . . . . . . .Assemblage 6, p. 14.


*Assemblage 1*

Heads discoid; pappus capillary (of soft hairs or bristles).

a. Receptacles chaffy (with small bracts that subtend the ovaries of the disk flowers); plants with woolly pubescence . . . . . . . . . . . . . . .*Filago*, p. 103.

aa. Receptacles bristly-hairy or naked.

b. Receptacles bristly-hairy.

c. Phyllaries hooked at the tip; stems and leaves without spines. . . . . . . .

. . . . . . . . . . . . . . . . . . . . . . . . . . . . . . . . . . . . . . . . . . . . . .*Arctium*, p. 207.

cc. Phyllaries not hooked at tip.

d. Pappus plumose; herbage spiny.

e. Filaments separate. . . . . . . . . . . . . . . . . . . . . . . . . . . . .*Cirsium*, p. 210.

ee. Filaments united, at least below . . . . . . . . . . . . . . . .*Silybum*, p. 219.

dd. Pappus various, of barbellate or capillary bristles, or of scales, or sometimes reduced or wanting.

    e. Phyllaries linear to ovate, tipped with short spines; stems and leaves spine-toothed . . . . . . . . . . . . . . . . . . . . . . . . . . . . .*Carduus*, p. 209.

    ee. Phyllaries tipped with pectinate, lacerate, fringed, or toothed appendages, or sometimes tipped with long spines; stems and leaves not spine-toothed . . . . . . . . . . . . . . . . . . . . . . . . . . . .*Centaurea*, p. 220.

bb. Receptacles naked or with but a few short hairs or bristles.

    c. Plants climbing and twining . . . . . . . . . . . . . . . . . . . . .*Mikania*, p. 33.

    cc. Plants not climbing or twining.

        d. Stems broadly winged, the wings spine-tipped; leaves spine-tipped; pappus barbellate; plants white-woolly . . . . . . . . .*Onopordum*, p. 218.

        dd. Stems not winged.

            e. Pappus double, the outer shorter; corollas purple . . .*Vernonia*, p. 16.

            ee. Pappus simple, all about the same length, sometimes scabrous.

                f. Pappus plumose; leaves often resin-dotted.

                    g. Heads in corymbiform clusters; corollas creamy-white . . . . . . . . . . . . . . . . . . . . . . . . . . . . . . . . . . . . . . . . . . . . . . . . . . . . . . . . .*Kuhnia*, p. 35.

                    gg. Heads in racemose or spicate clusters . . . . . . . . . . . .*Liatris*, p. 36.

                ff. Pappus neither plumose nor barbellate, often scabrous.

                    g. Middle or lower leaves opposite or whorled. . . . .*Eupatorium*, p. 21.

                    g. Middle leaves alternate or leaves all basal.

                        h. Phyllaries of about the same length and in essentially a single series.

                        i. Disk flowers purple . . . . . . . . . . . . . . . . . . . . . . .*Petasites*, p. 195.

                        ii. Disk flowers yellow, white, or orangish.

                            j. Leaves serrate, and triangular or hastate . . . . . .*Cacalia*, p. 196.

                            jj. Leaves various, not triangular or hastate.

                                k. Leaves palmately veined; phyllaries about 5 . . . . . . . . . . . . . . . . . . . . . . . . . . . . . . . . . . . . . . . . . . . . . . . . . . . . . . . . . . .*Cacalia*, p. 196.

                                kk. Leaves pinnately veined, never palmate; phyllaries more than 5 . . . . . . . . . . . . . . . . . . . . . . . . . . . . . . . . . . . . . . . . .

                                    l. Involucre about 5 mm high, cylindric. . . . . . .*Solidago*, p. 48.

                                    ll. Involucre 1.5 cm high, broader at base. . . .*Erechtites*, p. 196.

                    hh. Phyllaries in more than a single series and of unequal lengths.

                      i. Plants more or less white-woolly, at least on undersurface leaves.

                        j. Basal leaves prominent at anthesis, oblanceolate to obovate; cauline leaves few and reduced; dioecious; plants flowering in early spring. . . . . . . . . . . . . . . . . . . . . . . . . . .*Antennaria*, p. 104.

jj. Basal leaves absent at anthesis or basal and cauline leaves similar.

    k. Monoecious; heads alike, outer flowers carpellate, yellowish; inner flowers perfect; plants flowering in midsummer . . . . . . . . . . . . . . . . . . . . . . . . . . . . . . . . . . . . . . . . . . .*Gnaphalium*, p. 108.

    kk. Dioecious or essentially so; carpellate heads often with a few staminate flowers near the center; plants flowering in late summer . . . . . . . . . . . . . . . . . . . . . . . . . . . .*Anaphalis*, p. 108.

ii. Plants without white-woolly tomentum.

  j. Leaves triangular-hastate . . . . . . . . . . . . . . . . . .*Cacalia*, p. 196.

  jj. Leaves of various shapes, never triangular-hastate.

    k. Phyllaries in 1 or 2 series, linear; leaves usually less than 1 cm wide . . . . . . . . . . . . . . . . . . . . . . . . . . . . . . . . . . .*Erigeron*, p. 99.

    kk. Phyllaries imbricate in 2 or more series, broad; leaves usually wider than 1 cm, camphor-scented . . . . . . . .*Pluchea*, p. 103.


*Assemblage 2*

Heads discoid; pappus none, or if present, not capillary.

a. Carpellate heads of 1 or 2 flowers, enclosed in a hardened, spiny, or turberculate involucre; staminate heads on the same plant in terminal clusters.

  b. Involucre of carpellate heads ovoid to somewhat turbinate, bearing a single series of tubercules or spines or prickles near the apex; carpellate flowers solitary in head. . . . . . . . . . . . . . . . . . . . . . . .*Ambrosia*, p. 115.

  bb. Involucre of carpellate heads more oblong, enclosing a flower in each of two cavities and becoming a spiny bur with hooked prickles . . . . . . . . . . . . . . . . . . . . . . . . . . . . . . . . . . . . . . . . . . . . . . . . . . . . . . . . .*Xanthium*, p. 119.

aa. Carpellate heads of more than 1 or 2 flowers, not enclosed in a thick, hard involucre.

  b. Heads aggregated in clusters (glomerules), the clusters subtended by one or more foliaceous bracts . . . . . . . . . . . . . . . . . . . . .*Elephantopus*, p. 20.

  bb. Heads not in glomerules.

    c. Pappus of 2–6 rigid awns, retrorsely barbed; phyllaries in 2 series, often foliaceous . . . . . . . . . . . . . . . . . . . . . . . . . . . . . . . . . .*Bidens*, p. 166.

    cc. Pappus none or at most a short crown or cup.

      d. Phyllaries multiseriate, spiny or variously appendaged (pectinate, lacerate, fringed, or toothed), margins hyaline. . . . . . .*Centaurea*, p. 220.

      dd. Phyllaries sometimes multiseriate and imbricate but not appendaged as above.

        e. Leaves, at least the middle and lower, opposite.

 f. Leaves entire, unlobed, crenate, or dentate; heads of greenish-white flowers, small, barely 5 mm high..................*Iva*, p. 113.

 ff. Leaves large, lobed, viscid-pubescent, usually with axillary stipules; heads larger than 5 mm.....................*Polymnia*, p. 121.

ee. Leaves alternate, variously dissected, often aromatic; heads in corymbs.

 f. Receptacles strongly conic..................*Matricaria*, p. 183.

 ff. Receptacles flat or convex.

  g. Heads small, less than 5 mm wide...........*Artemisia*, p. 188.

  gg. Heads wider than 5 mm.

   h. Leaves with crenate margins, often lobed at base ...........................................*Chrysanthemum*, p. 185.

   hh. Leaves 2–3 times pinnately divided or dissected, strongly scented ....................................*Tanacetum*, p. 188.


*Assemblage 3*

Heads radiate; pappus capillary (of soft hairs).

a. Ray flowers yellow or orange.

 b. Phyllaries multiseriate, unequal, imbricate.

  c. Heads large, 3–5 cm wide; leaves irregularly dentate, white-woolly or velvety beneath; pappus single.....................*Inula*, p. 110.

  cc. Heads to 2.5 cm wide.

   d. Phyllaries bearing 3–7 conspicuous glands; leaves opposite, dissected into linear lobes; pappus of 8–15 scales dissected into soft bristles ...........................................*Dyssodia*, p. 179.

   dd. Phyllaries not gland-dotted.

    e. Heads 1–2.5 cm wide.

     f. Lower leaves linear, the veins appearing parallel....*Pityopsis*, p. 45.

     ff. Lower leaves oblanceolate to ovate or oblong, pinnately veined.

      g. Pubescence of lower stem and leaves of fine flagelliform hairs....................................................*Chrysopsis*, p. 46.

      gg. Pubescence of lower stem and leaves of rigid hairs....................................................*Heterotheca*, p. 46.

    ee. Heads less than 1 cm wide.....................*Solidago*, p. 48.

 bb. Phyllaries in a single series, nearly equal but sometimes with a few shorter ones at base.

  c. Heads solitary on scaly scapes, plants flowering in early spring before leaves appear................................*Tussilago*, p. 194.

  cc. Heads many, not solitary; stems leafy; plants flowering after leaves appear........................................*Senecio*, p. 198.

aa. Ray flowers variously colored, not yellow or orange.

  b. Ray flowers few, not more than 5.

    c. Heads in a corymb; achenes silky....................*Aster,* p. 67.

    cc. Heads in a narrow, elongated panicle .............*Solidago,* p. 48.

  bb. Ray flowers many, more than 8.

    c. Phyllaries imbricate in 3–5 series ...................*Aster,* p. 67.

    cc. Phyllaries about equal, or if imbricated, ray flowers hardly longer than the disk flowers.

      d. Heads small, 5 mm or less wide; disk flowers about 20 ...........
.........................................*Conyza,* p. 96.

      dd. Heads at least 1 cm wide; disk flowers more than 20.............
.........................................*Erigeron,* p. 99.

## Assemblage 4

Heads radiate; pappus a circle of awns, of chaffy or rigid bristles, or of scales sometimes dissected into bristles, often deciduous.

a. Plants gland-dotted; rays yellow or orange, sometimes purplish at base.

  b. Leaves opposite, dissected into linear lobes; phyllaries bearing conspicuous glands; pappus of 8–15 scales dissected into 5–10 soft bristles .....
.........................................*Dyssodia,* p. 179.

  bb. Leaves alternate or all basal.

    c. Leaves basal, linear; pappus of very thin scales....*Hymenoxys,* p. 174.

    cc. Leaves cauline, not basal.

      d. Pappus of awns or linear scales, quickly deciduous; leaves resin-dotted, often gummy, not decurrent; receptacle flat or somewhat convex.

        e. Heads 1 cm or more wide.....................*Grindelia,* p. 43.

        ee. Heads less than 1 cm wide .................*Gutierrezia,* p. 44.

      dd. Pappus of several awn-tipped, thin hyaline scales; leaves often decurrent, not gummy; receptacle convex to conic .....*Helenium,* p. 177.

aa. Plants not gland-dotted; rays white, pink, or blue.

  b. Leaves opposite, ovate, serrate, petiolate ..........*Galinsoga,* p. 172.

  bb. Leaves alternate, linear to narrowly elliptic, entire, sessile ...........
.........................................*Boltonia,* p. 65.

## Assemblage 5

Heads radiate; pappus none, or a short crown or ridge, or of a few teeth, awns, or scales that correspond to the angles of the achenes, sometimes with additional tiny scales or bristles.

a. Receptacles naked.

  b. Leaves simple.

    c. Leaves all basal; heads scapose; pappus none ..........*Bellis,* p. 65.

cc. Stems leafy; leaves narrow, mostly entire; pappus of short scales or bristles and 2–4 long awns........................*Boltonia*, p. 65.

bb. Leaves compound, dissected, laciniate-toothed, or lobed.

  c. Plants annual; receptacles conic to hemispheric; leaves with filiform to linear segments ...........................*Matricaria*, p. 183.

  cc. Plants perennial; receptacles flat or slightly convex; leaves with broader segments ............................*Chrysanthemum*, p. 185.

aa. Receptacles chaffy.

  b. Ray flowers either white to pink or purple to reddish-purple.

    c. Ray flowers reddish-purple to light purple, rarely white .............
.......................................*Echinacea*, p. 139.

    cc. Ray flowers white or pinkish.

      d. Leaves opposite, linear, or lanceolate .............*Eclipta*, p. 133.

      dd. Leaves alternate.

        e. Leaves finely dissected into pinnate, often filiform, segments.

          f. Rays white or pink, disk corollas white..........*Achillea*, p. 180.

          ff. Rays white, disk corollas yellow...............*Anthemis*, p. 181.

        ee. Leaves not finely dissected, but sometimes pinnatifid or bipinnatifid.

          f. Leaves with 20 or more dentate or crenate teeth on each side of blade, or the leaves 1–2 times pinnatifid.......*Parthenium*, p. 128.

          ff. Leaves with about 10 or fewer shallow teeth on each side of blade
.......................................*Verbesina*, p. 157.

  bb. Ray flowers yellow to orange, sometimes brown or purple at the base.

    c. Stems winged by decurrent leaf bases .............*Verbesina*, p. 157.

    cc. Stems without wings.

      d. Receptacles flat or convex.

        e. Phyllaries in more than two series.

          f. Achenes winged, formed only in ray flowers ......*Silphium*, p. 123.

          ff. Achenes not winged, formed only in disk flowers ..............
.......................................*Helianthus*, p. 140.

        ee. Phyllaries in one or two series.

          f. Phyllaries in one series .....................*Polymnia*, p. 121.

          ff. Phyllaries in two series.

            g. Pappus absent, or of two short crowns or ridges, not rigid.

              h. Ray flowers usually 5 .................*Chrysogonum*, p. 127.

              hh. Ray flowers more than 5 .................*Coreopsis*, p. 158.

            gg. Pappus of rigid, often barbed, awns.

              h. Achenes terete; plants aquatic..........*Megalodonta*, p. 164.

              hh. Achenes flattened or quadrangular; plants terrestrial..........
.......................................*Bidens*, p. 166.

dd. Receptacles conic or columnar.
 e. Leaves pinnately divided into narrow segments; leaves often white-tomentose; phyllaries with thin, scarious margins . . . . . . . . . . . . . . .
  . . . . . . . . . . . . . . . . . . . . . . . . . . . . . . . . . . . . . . . . . . *Anthemis*, p. 181.
ee. Leaves, if pinnate, not divided into narrow segments.
 f. Leaves opposite . . . . . . . . . . . . . . . . . . . . . . . . . . . *Heliopsis*, p. 130.
ff. Leaves alternate.
 g. Achenes flattened to weakly 4-sided, margined or winged; leaves pinnatifid; disk green . . . . . . . . . . . . . . . . . . . . . . .*Ratibida*, p. 140.
gg. Achenes 4-sided, marginless; leaves unlobed, 3-lobed to trifolio-late, or rarely pinnatifid; disk purple, brown, or yellowish-green . . . . . . . . . . . . . . . . . . . . . . . . . . . . . . . . . . . . . . . .*Rudbeckia*, p. 134.

*Assemblage 6*

Heads ligulate (Tribe X. Lactuceae).
a. Pappus absent; plants annual.
 b. Cauline leaves present; peduncles not inflated. . . . . . . .*Lapsana*, p. 225.
 bb. Leaves all basal or cauline leaves reduced to mere bracts; peduncles infla-ted . . . . . . . . . . . . . . . . . . . . . . . . . . . . . . . . . . . . . . .*Arnoseris*, p. 225.
aa. Pappus present; plants annual or perennial.
 b. Pappus chaffy, or of chaff and bristles.
 c. Involucre of 2 series, the inner of 8–10 phyllaries, the outer of 5 short phyllaries; pappus of uniform narrow scales; flowers blue, rarely white or pink . . . . . . . . . . . . . . . . . . . . . . . . . . . . . . . . . . . . . .*Cichorium*, p. 225.
 cc. Involucre of a single series; pappus of scales alternating with bristles; flowers yellow to orange . . . . . . . . . . . . . . . . . . . . . . . . . . .*Krigia*, p. 228.
 bb. Pappus plumose or capillary.
 c. Pappus bristles of most flowers plumose or barbellate.
  d. Leaf blades linear-lanceolate, entire . . . . . . . . . . . .*Tragopogon*, p. 233.
  dd. Leaf blades toothed or lobed.
   e. Cauline leaves present. . . . . . . . . . . . . . . . . . . . . . . . .*Picris*, p. 232.
   ee. Leaves all basal or crowded near the base of the stem.
    f. Receptacles naked or with a few hairs. . . . . . . . .*Leontodon*, p. 231.
    ff. Receptacles chaffy . . . . . . . . . . . . . . . . . . . . . .*Hypochaeris*, p. 229.
 cc. Pappus simple, not plumose, some white scales occasionally present.
  d. Achenes fusiform with slender beaks, the surface with sharp points; scapose herbs with hollow stems; heads solitary. . .*Taraxacum*, p. 236.
  dd. Achenes flat or columnar, the surface without sharp points, not muri-cate.
   e. Achenes flat or flattish.

f. Achenes beaked or at least with a thickened summit; flowers fewer than 50 per head; involucre cylindric or conic; flowers blue, purple, or yellow ................................*Lactuca*, p. 240.

ff. Achenes without beak; flowers more than 50 per head; involucre ovoid or campanulate; flowers yellow ...........*Sonchus*, p. 236.

ee. Achenes columnar, not or only slightly flattened.

f. Corollas yellow or orange.

g. Achenes markedly narrowed at apex; phyllaries in two series, the outer much reduced; pappus white.............*Crepis*, p. 248.

gg. Achenes truncate and little narrowed at apex; phyllaries in three series; pappus gray, tawny, or brown .........*Hieracium*, p. 254.

ff. Corollas white, cream, or pink; heads drooping. .*Prenanthes*, p. 250.

# SYSTEMATIC TEXT

The descriptions of genera and species are intended for the separation of those taxa occurring in Ohio. They may or may not be adequate in separating species of other political subdivisions.

Specific names of native species are printed in **boldface.** Names of species that are nonindigenous, whether adventive or naturalized, are printed in SMALL CAPITALS and in most instances are mapped because they are indeed a part of the flora of Ohio. Flowering periods are indicated at the ends of the species descriptions.

Synonyms are included when appropriate, usually in those cases in which the names used here differ from those used by Fernald (1950) and Cronquist (1952). In addition, synonyms from more recent monographs or revisions are listed.

## Asteraceae (Compositae). SUNFLOWER FAMILY
### TRIBE I. VERNONIEAE CASS.

a. Heads not in glomerules, many-flowered; phyllaries uniform and obviously imbricate; pappus in a double row . . . . . . . . . . . . . . . . . . . . . .1. *Vernonia*

aa. Heads in small clusters (glomerules), each head with few (2–5) flowers, outer phyllaries leafy, the inner narrow, flattened and dry; pappus in a single row of chaffy bristles . . . . . . . . . . . . . . . . . . . . . .2. *Elephantopus*

### 1. Vernonia Schreb. IRONWEED

Heads discoid, in corymb-like cymes. Flowers perfect and fertile. Phyllaries imbricated in several series, persistent and spreading. Receptacle flat, naked. Corollas dark purple, rarely white, regular. Pappus in two series, the outer shorter and of bristles or scales, the inner of capillary bristles. Achenes ribbed, glabrous or somewhat pubescent. Perennial herbs. Leaves alternate, lanceolate, serrate, sessile to subsessile. The species of this genus, at least in North America, are likely to hybridize freely when two or more occur sympatrically. For detailed treatment see Jones and Faust (1978), Faust (1972), and Jones (1972).

a. Phyllaries prolonged into threadlike filiform tips. . . . .1. *V. noveboracensis*

aa. Phyllaries obtuse to acuminate, not filiform-tipped.

b. Lower leaf surfaces tomentulose, hairs often crooked; heads with more than 30 flowers . . . . . . . . . . . . . . . . . . . . . . . . . . . . . . . . . . .2. *V. missurica*

bb. Lower leaf surfaces glabrous or minutely short-pubescent; heads small, with fewer than 30 flowers.

c. Leaf surfaces, especially the lower, conspicuously punctate; branches of the inflorescence forming a crowded or compact cyme . . . . . . . . . . . . . . . . . . . . . . . . . . . . . . . . . . . . . . . . . . . . . . . . . . . . . . .3. *V. fasciculata*

cc. Leaf surfaces only remotely punctate, if at all; branches of the inflorescence open and spreading. . . . . . . . . . . . . . . . . . . . . . . . .4. *V. gigantea*

1. **Vernonia noveboracensis** (L.) Michx.   NEW YORK IRONWEED

Stems to 2 m tall, glabrous or sparsely pubescent; leaves long-lanceolate to linear-lanceolate, acuminate, serrate, the lower surface scantly pubescent with short hairs; heads arranged in an open cyme; heads 30–45-flowered; phyllaries

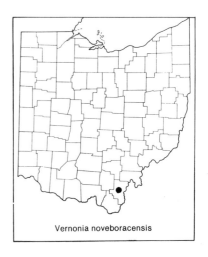

Vernonia noveboracensis

× 2

× ½

Vernonia noveboracensis

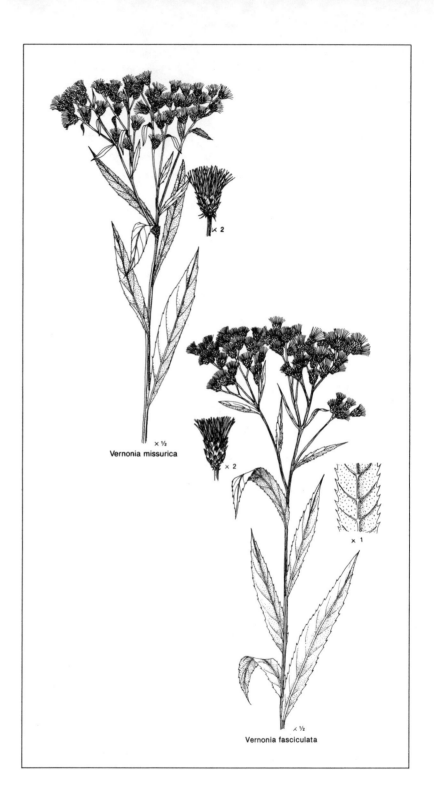

x 2

x ½
Vernonia missurica

x 2

x 1

x ½
Vernonia fasciculata

purplish, tips loosely ascending or recurved, filiform; pappus purple; achenes pubescent on angles. Moist, open areas. Aug.–Oct.

Although this species is represented in Ohio by a single specimen collected in Gallia County, with its phyllaries prolonged into filiform tips, there is no doubt as to its identity.

**2. Vernonia missurica** Raf.    Missouri Ironweed

Stems to 2 m tall, pubescent or merely tomentulose; leaves lanceolate, tomentose beneath, especially along the veins, the hairs often crooked, inflorescence cymose, widely branched; heads 30–45-flowered; phyllaries purple, their tips rounded to mucronate; pappus purplish, becoming tawny with age. Open ground. July–Sept.

According to Jones and Faust (1978), this species does not occur in Ohio, but several specimens from northwestern and central Ohio counties, which appear to be correctly identified as members of this species, were probably not known to them. There are also several specimens in The Ohio State University Herbarium that clearly represent intergrades between this species and *V. gigantea*.

**3. Vernonia fasciculata** Michx.    Western Ironweed

Stems to 2 m tall, glabrous; leaves numerous, ascending, lanceolate, glabrous beneath, somewhat scabrous above, conspicuously punctate beneath; inflorescence usually very compact, flat; heads 10–26-flowered, crowded; phyllaries rounded to acute, entire or ciliate; achenes glabrous; pappus purple to tawny. Known from wet prairie sites in a few counties. July–Sept.

Our specimens are representative of *ssp. fasciculata*. Several specimens examined suggest hybridization with *V. gigantea*.

**4. Vernonia gigantea** (Walt.) Trel. ex Branner & Coville    Tall Ironweed
   *V. altissima* Nutt.

Stems to 2 m tall, usually glabrate below, puberulent above; leaves linear to linear-lanceolate, glabrous to slightly scabrous above, pubescent on veins beneath, nearly always serrate; inflorescence loose, open, rather flat; phyllaries imbricate, usually purplish, ciliate, their tips obtuse or mucronate, especially the inner; achenes pubescent on ribs; pappus tan to brown. Common and widely distributed. Moist fields. Aug.–Oct.

This species is best treated as consisting of two subspecies, of which only ssp. *gigantea* occurs in Ohio. Intergrades with *V. noveboracensis* and *V. missurica* occur in some areas. For further study of taxonomy, see Urbatsch (1972).

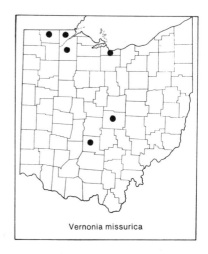

Vernonia missurica

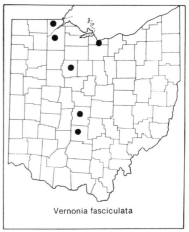

Vernonia fasciculata

Vernonia gigantea

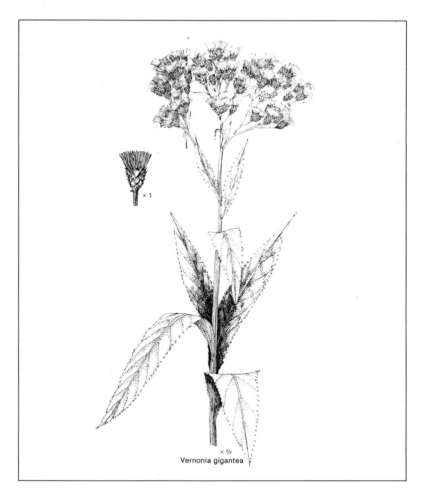

× 1

× ½
Vernonia gigantea

## 2. Elephantopus L.  ELEPHANT'S-FOOT

Heads discoid, of 1–5 perfect and fertile flowers, arranged in clusters or glomerules subtended by foliaceous bracts, in all suggesting a single head. Receptacle flattish, naked. Corollas blue-purple, somewhat unevenly cleft. Pappus of scales tapering to terminal bristles. Achenes truncate, about 10-ribbed. Branched perennial herbs, with stems tall and soft-pubescent. Leaves cauline, alternate, serrate to merely crenate, elliptic to obovate, tapering at base to a winged petiole, pilose beneath, glabrous above or with a few scattered hairs.

Primarily tropical and subtropical in distribution, the genus is represented in Ohio by a single species. See Clonts and McDaniel (1978) for a taxonomic treatment of the genus in North America.

1. **Elephantopus carolinianus** Raeuschel  ELEPHANT'S-FOOT
Open woods and thickets, primarily in south-central Ohio. Aug.–Oct.

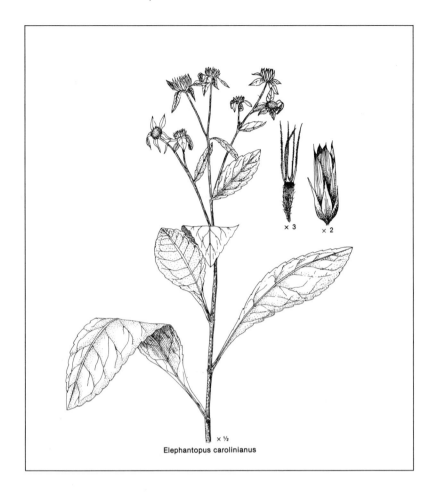

×3  ×2

× ½

Elephantopus carolinianus

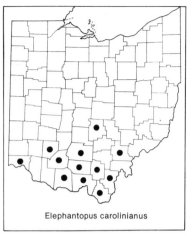

Elephantopus carolinianus

## TRIBE II. EUPATORIEAE CASS.

a.  Pappus bristles capillary.
  b.  Involucre of more than 4 phyllaries ................3. *Eupatorium*
  bb. Involucre of 4 phyllaries ...........................4. *Mikania*
aa. Pappus bristles plumose or barbellate.
  b.  Heads in corymbiform clusters; corollas creamy-white .......5. *Kuhnia*
  bb. Heads in racemes or spikes; corollas rose to purple, rarely white.......
    ..............................................6. *Liatris*

### 3. Eupatorium L.  BONESET

Heads discoid, in a corymbose inflorescence. Flowers tubular and perfect. Receptacle flat or conic, naked. Phyllaries usually in several series, imbricated. Corollas pink, purple, or rarely white. Achenes ribbed, glabrous to slightly pubescent. Pappus in one or two series of capillary bristles, the outer occasionally intermixed with scales. Perennials. Leaves entire or toothed, alternate, opposite, or whorled.

A large cosmopolitan genus centered in the American tropics. In a series of papers, King and Robinson (see references in King and Robinson, 1970) reclassified this large and cumbersome genus into more natural groupings by moving many species from *Eupatorium* to several segregate genera. Especially noteworthy are two species occurring in the Ohio flora which have been transferred to the genus *Ageratina*. Although the revision of the genus *Ageratina* by Clewell and Wooten (1971) appears to justify this change, I have elected to retain these species within the genus *Eupatorium*, pending further study and clarification of this rather confusing taxonomic situation.

a. Leaves whorled, or if opposite less than 1.0 cm wide.
  b. Corollas white; leaves linear, entire to laciniate-toothed, and with conspicuous axillary fascicles . . . . . . . . . . . . . . . . . . . . . .1. *E. hyssopifolium*
  bb. Corollas purple to pale pinkish-purple.
    c. Inflorescence a nearly flat-topped corymb; heads generally of more than 8 flowers; stems glaucous, purple throughout or dotted with purple . . . . . . . . . . . . . . . . . . . . . . . . . . . . . . . . . . . . . . . . . . . . .2. *E. maculatum*
    cc. Inflorescence convex or rounded at top; flowers generally fewer than 8 per head.
      d. Stems hollow, usually glaucous, usually purple throughout . . . . . . . . . . . . . . . . . . . . . . . . . . . . . . . . . . . . . . . . . . . . . . .3. *E. fistulosum*
      dd. Stems solid, somewhat glaucous, often purplish at nodes . . . . . . . . . . . . . . . . . . . . . . . . . . . . . . . . . . . . . . . . . . . . . . . . .4. *E. purpureum*
aa. Leaves opposite or occasionally alternate above.
  b. Phyllaries mucronate, their conspicuously long, scarious margins with dark sessile glands which often extend onto peduncles . . . .5. *E. album*
  bb. Phyllaries without conspicuous scarious margins and mucronate tips.
    c. Leaves sessile, subsessile, or connate-perfoliate.
      d. Leaves connate-perfoliate . . . . . . . . . . . . . . . . . . . . . .6. *E. perfoliatum*
      dd. Leaves not connate-perfoliate.
      e. Leaves rarely more than 1.0 cm wide, usually whorled, often bearing fascicles in axils . . . . . . . . . . . . . . . . . . . . . . . . . . . .1. *E. hyssopifolium*
      ee. Leaves considerably wider than 1.0 cm.
        f. Leaf blades narrowed to the base, lanceolate, acuminate, usually serrate only beyond the middle; stems soft-pubescent . . . . . . . . . . . . . . . . . . . . . . . . . . . . . . . . . . . . . . . . . . . . . . . . . . .7. *E. altissimum*
        ff. Leaf bases rounded or truncate.

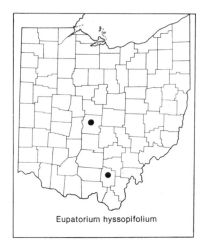

Eupatorium hyssopifolium

g. Stems glabrous below the inflorescence; leaf blades lanceolate, acuminate, rounded at the base .............8. *E. sessilifolium*

gg. Stems pubescent; leaf blades more ovate than lanceolate, rather broad and truncate at base ...............9. *E. rotundifolium*

cc. Leaves definitely petiolate.

d. Flowers blue, violet, or pinkish-purple.

e. Flowers pinkish-purple when fresh and when dry, fewer than 25 per head; receptacle flat........................10. *E. incarnatum*

ee. Flowers blue or violet when fresh, pinkish to purple when dry, more than 25 per head; receptacle conic............11. *E. coelestinum*

dd. Flowers white.

e. Plants much-branched above; mature inflorescence 30 cm or more wide at the crown..........................12. *E. serotinum*

ee. Plants slender, with few branches above; mature inflorescence much narrower than 30 cm.

f. Leaves small, crenate-serrate, less than 8.0 cm long, rounded to truncate at base; petioles less than 2.0 cm long .....13. *E. aromaticum*

ff. Leaves larger, sharply serrate, exceeding 8.0 cm in length; petioles generally longer than 2.0 cm..................14. *E. rugosum*

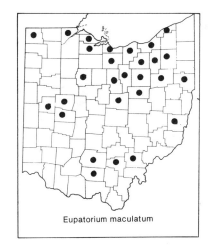

Eupatorium maculatum

## 1. **Eupatorium hyssopifolium** L.

Stems strigose or evenly pubescent; leaves verticillate, usually in 4s or sometimes 2s and hence opposite, linear or linear-lanceolate, mostly entire, usually trinerved, glabrous or glandular-punctate, bearing fascicles in axils; phyllaries imbricate, rounded to slightly acute; flowers usually 5 per head, white. Fields and open areas, rare in Ohio. Aug.–Oct.

Two varieties may be recognized in this species, but only var. *laciniatum* has been observed in Ohio.

## 2. **Eupatorium maculatum** L.   Spotted Joe-Pye Weed

Stems often glaucous, evenly purple or more often spotted; leaves whorled in 4s or 5s, lanceolate or lance-ovate, serrate; inflorescence nearly always flat-topped; phyllaries imbricate, obtuse; flowers purple or pinkish-purple, more than 8 per head. Scattered distribution in Ohio, occurring in moist habitats. July–Sept.

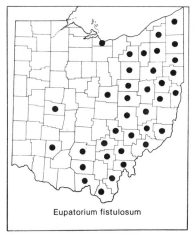

Eupatorium fistulosum

## 3. **Eupatorium fistulosum** Barratt   Joe-Pye Weed

Stems usually glaucous, most often purple, hollow; leaves whorled in 4s and 5s, bluntly serrate; inflorescence convex or rounded; flowers fewer than 8 per head. Well-represented in bottomlands and moist thickets, especially in the unglaciated regions. July–Sept.

## 4. **Eupatorium purpureum** L.   Purple Joe-Pye weed
*Eupatorium falcatum* Michx.

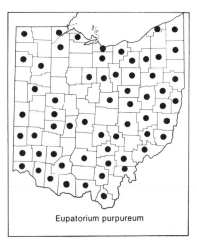

Eupatorium purpureum

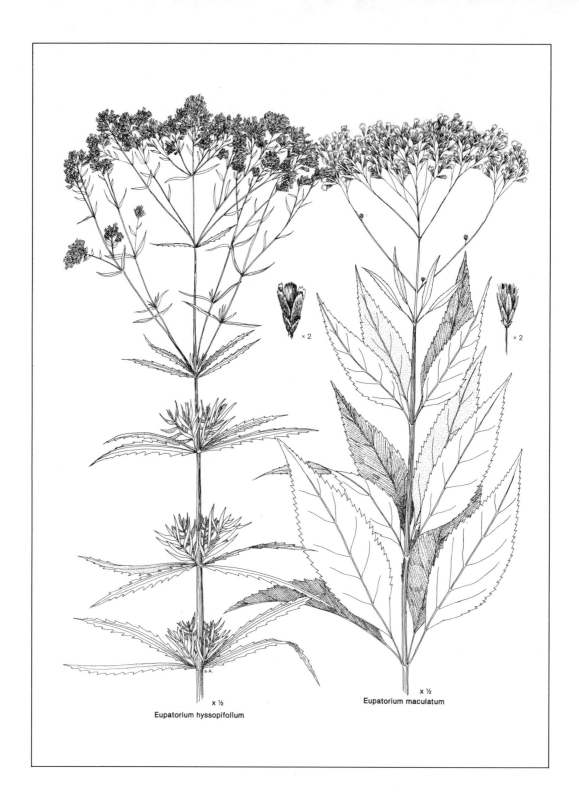

x ½

Eupatorium hyssopifolium

x 2

x 2

x ½

Eupatorium maculatum

x ½
Eupatorium fistulosum

x ½
Eupatorium purpureum

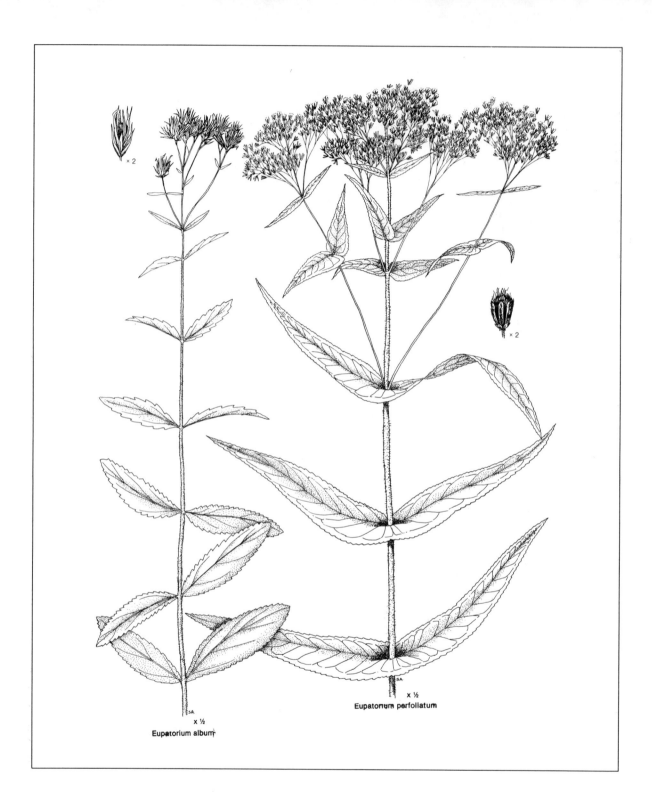

x 2

x 2

x ½
Eupatorium perfoliatum

x ½
Eupatorium album

Stems usually less glaucous than those of *E. fistulosum*, usually purplish only at the nodes, pith solid; leaves whorled in 3s and 4s, or 5s, lanceolate to ovate-lanceolate, sharply serrate-dentate; inflorescence convex or rounded; flowers fewer than 8 per head, pink or pinkish-purple. Well represented in upland woods and fields. July–Sept.

5. **Eupatorium album** L.   White Thoroughwort
    incl. var. *glandulosum* (Michx.) DC.
Stems villous-pubescent; leaves opposite, lanceolate to lance-ovate, essentially sessile, glandular-punctate; phyllaries narrow, long-acuminate, mucronate, often with sessile glands; flowers 5 per head, white. Open woodlands. July–Oct.

6. **Eupatorium perfoliatum** L.   Common Boneset
Stems villous with spreading hairs; leaves opposite, connate-perfoliate, crenate-serrate; inflorescence flat-topped, many-branched; phyllaries imbricate, pubescent, flowers 10 or more per head, white. Common in low, wet ground along stream banks and river bottoms. July–Oct.

Several varieties of dubious merit have been described, but they are hardly discernable.

7. **Eupatorium altissimum** L.   Tall Thoroughwort
    *Ageratina altissima* (L.) King & Robinson
Stems soft-pubescent throughout; leaves opposite, trinerved, glandular-punctate, sessile, lanceolate, serrate beyond middle; phyllaries rounded; flowers about 5 per head, white. Rather common in woodlands and forest borders. Aug.–Sept.

8. **Eupatorium sessilifolium** L.   Upland Boneset
Stems with short spreading pubescence in the inflorescence, otherwise glabrous; leaves opposite, sessile, broadly rounded at base, lanceolate, serrate,

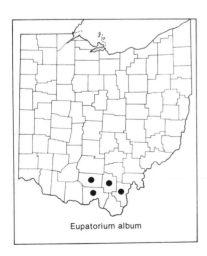

Eupatorium album

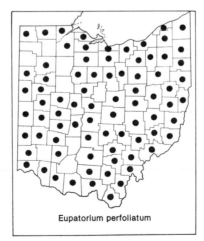

Eupatorium perfoliatum

× 1

S A
x ½

Eupatorium altissimum

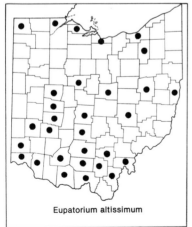

Eupatorium altissimum

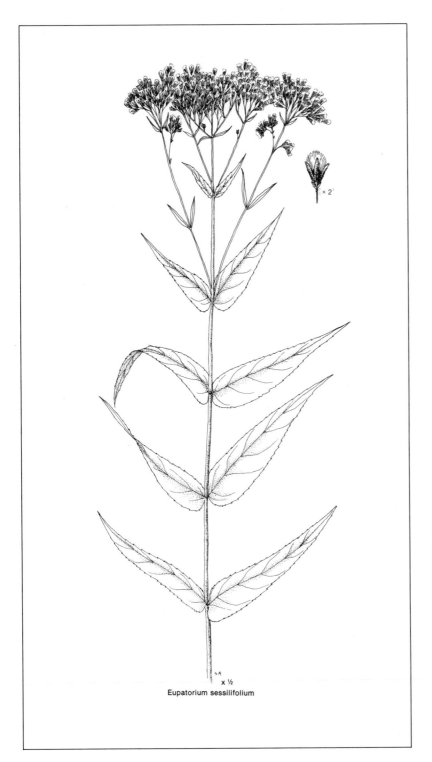

×2'

×½
Eupatorium sessilifolium

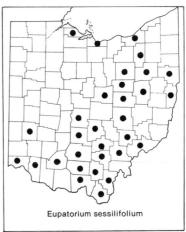

Eupatorium sessilifolium

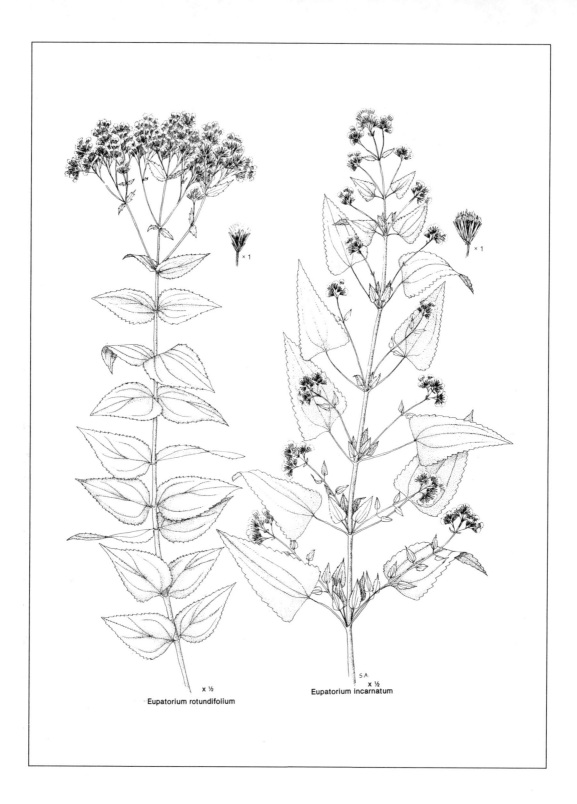

x ½
Eupatorium rotundifolium

x ½
Eupatorium incarnatum

long-acuminate; phyllaries imbricate, rounded, villous; flowers about 5 per head, white. Common on dry slopes and in woodlands, mostly in the southern and southeastern sections of Ohio. July–Sept.

### 9. **Eupatorium rotundifolium** L.   ROUNDLEAF THOROUGHWORT

Stems pubescent with short, spreading hairs, especially above; leaves opposite, sessile or subsessile, mostly crenate or crenate-serrate, ovate or ovate-lanceolate, rounded or truncate at base; phyllaries linear to linear-lanceolate, acute; flowers about 5 per head, white. Uncommon; it occurs on dry road-banks and in dry woodlands. July–Sept.

Two weakly separated varieties may be distinguished: var. *rotundifolium*, which often has more crenate leaf margins and lateral veins that arise at about the junction of the blade and petiole; and var. *ovatum* (Bigel.) Torr. (illustrated), which usually has nearly serrate leaf margins and lateral veins that arise above the junction of the blade and the petiole. For a study of the *E. rotundifolium* complex, see Montgomery and Fairbrothers (1970).

### 10. **Eupatorium incarnatum** Walt.   PINK THOROUGHWORT

Stems weak, semi-prostrate, free-branching, scarcely puberulent; leaves opposite, deltoid, petiolate, crenate to somewhat serrate; inflorescences borne terminally at tips of several lateral branches; phyllaries linear, each with 2–4 white nerves; receptacles flat, flowers 15–25 per head, pink or pinkish. Rare, specimens from only two southern counties, although apparently frequent locally; at woodland borders and field margins. Aug.–Sept.

### 11. **Eupatorium coelestinum** L.   MIST-FLOWER. AGERATUM

    *Conoclinium coelestinum* (L.) DC.

Stems tall, puberulent; leaves opposite, petiolate, crenate or irregularly serrate, deltoid or ovate; phyllaries linear, firm, long-pointed; receptacles conic; flowers 30 or more per head, blue or violet and turning purplish when dry. Woodlands, stream margins, and fields. Known from counties in southern and east-central Ohio. July–Oct.

### 12. **Eupatorium serotinum** Michx.   LATE-FLOWERING THOROUGHWORT

Stems puberulent above, sparsely so below, freely branched in inflorescence; leaves opposite, lanceolate or ovate-lanceolate, petiolate, coarsely serrate; phyllaries somewhat rounded to obtuse, pubescent; flowers 10–15 per head, white. Rather common in the southern part of Ohio, less so northward. The species occurs primarily on flood plains and in moist woods. Aug.–Oct.

This species is often difficult to distinguish from *E. rugosum*. It differs chiefly in having much more branched, and therefore wider, inflorescences, and in having more lanceolate leaves.

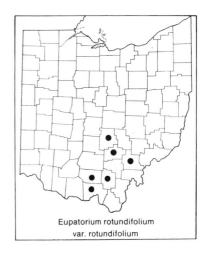

Eupatorium rotundifolium
var. rotundifolium

Eupatorium rotundifolium
var. ovatum

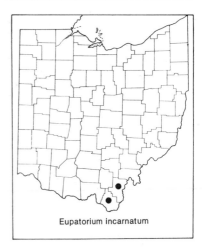

Eupatorium incarnatum

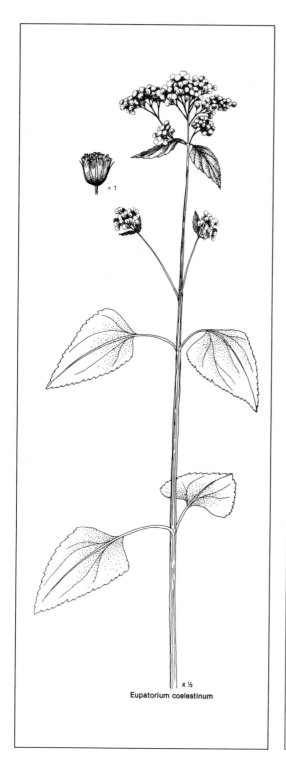

×1

× ½

Eupatorium coelestinum

× 1

× ½

Eupatorium serotinum

13. **Eupatorium aromaticum** L.  SMALL WHITE SNAKEROOT

    *Ageratina aromatica* (L.) Spach

    *Ageratina aromatica* (L.) King & Robinson

Stems minutely pubescent in the inflorescence, otherwise glabrous; leaves opposite, thickish, generally short-petiolate, crenate-serrate; phyllaries rounded to acute; flowers usually more than 9 per head, white. Open areas and thickets. Aug.–Oct.

    This species is similar to *E. rugosum*, but is generally smaller, generally shorter petioled, and has mostly crenate leaves. It is known to intergrade with *E. altissimum* (Clewell and Wooten, 1971).

14. **Eupatorium rugosum** Houtt.  WHITE SNAKEROOT

    *E. urticaefolium* Reich.

Stems tall, glabrous and often glaucous below the inflorescence; leaves opposite, long-petiolate, broadly ovate to ovate-lanceolate, sharply serrate; phyllaries glabrous to short-hairy, obtuse, 2–3-nerved; flowers 12–25 per head, white. Common in woodlands and fields. July–Oct.

    Plants of this species are poisonous to livestock, and the toxic content is transmissible to humans through milk from cows. Reports of livestock deaths from eating these plants occur nearly every year.

### 4. Mikania Willd.  CLIMBING HEMPWEED

Heads discoid, small, many, arranged in corymbiform clusters, each head 4-flowered, the flowers perfect. Phyllaries 4, equal. Receptacle naked. Corollas tubular, whitish to lilac or purplish. Achenes 5-angled. Pappus of a single series of capillary bristles. Perennial, climbing or twining vines. Leaves opposite, simple, petiolate, triangular-cordate to hastate.

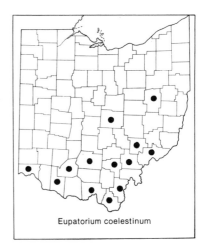

Eupatorium coelestinum

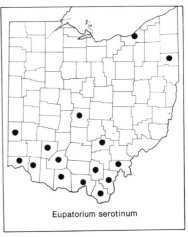

Eupatorium serotinum

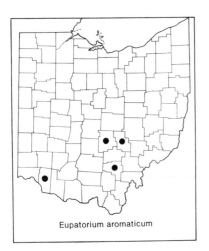

Eupatorium aromaticum

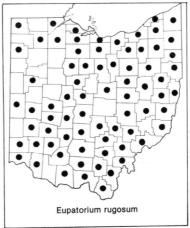

Eupatorium rugosum

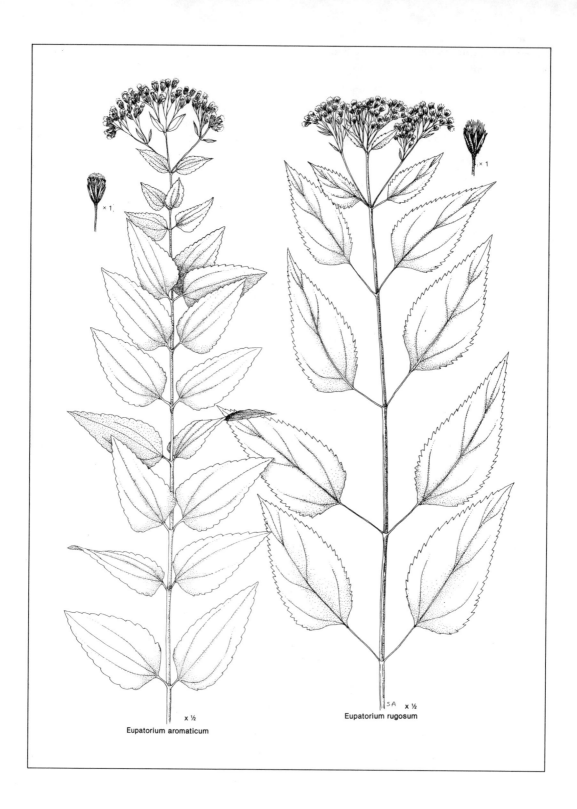

x ½

Eupatorium aromaticum

x ½

Eupatorium rugosum

1. **Mikania scandens** (L.) Willd.

Reluctantly listed here, based on two specimens collected in Hamilton County by R. Buchanan in the 19th century and deposited in the Curtis Gates Lloyd Herbarium of the University of Cincinnati. However, this southern species might be found again in Ohio, in moist areas of flood plains and swamps (not illustrated or mapped). July–Oct.

## 5. Kuhnia L.  FALSE BONESET

Heads discoid, in terminal corymbiform clusters. Phyllaries imbricate, the outer shorter. Flowers creamy-white, perfect and fertile. Achenes linear, puberulent. Pappus of 20 long bristles. Stems densely puberulent to subglabrous. Leaves linear to ovate, margins inrolled, entire to evenly toothed, mostly sessile, gland-dotted beneath.

A small genus of 6 species, native to North America and Mexico. In Ohio, the genus is represented by one species consisting of two weakly separated varieties.

× 3

× 2

× ½
Kuhnia eupatorioides

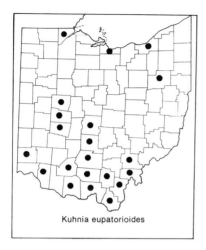

Kuhnia eupatorioides

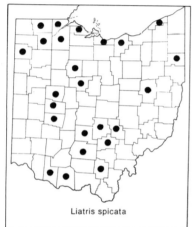

Liatris spicata

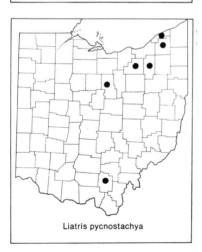

Liatris pycnostachya

1. **Kuhnia eupatorioides** L.   FALSE BONESET
*Brickellia eupatorioides* (L.) Shinners
Open habitats in fields and prairies, especially in slightly sandy soils. Midsummer–late fall.

Most previous treatments (Cronquist, 1952; Fernald, 1950) recognize two varieties in Ohio. Critical examination of herbarium and field material reveals that var. *eupatorioides* passes into var. *corymbulosa* T. & G. and that no combination of characters consistently separates the two. For this reason, no detailed separation is employed in this treatment.


6. Liatris Schreb.   B L A Z I N G - S T A R
Heads discoid, in spikes or short racemes. Phyllaries imbricate in several series. Receptacle flat, naked. Flowers perfect, seed-forming, rose-pink, purple, or white. Pappus of barbellate or plumose capillary bristles. Achenes ribbed, slender. Perennial herbs. Leaves alternate, entire, often punctate.

The genus consists of about thirty species, native to the north temperate region of North America. Seven species are known from Ohio, although one is regarded as a transient, as its normal range is farther west.

The most detailed taxonomic study of this genus is that of Gaiser (1946). Most species in our range are easily separated, but some, namely *L. scariosa* and *L. aspera*, are sometimes troublesome. I do not consider *L. novae-angliae* (Lunell) Shinners or *L. scabra* (Greene) K. Schum. to be present in Ohio, at least from the material I have examined during this study. The few specimens I have examined that approach *L. scabra* may be hybrids, either with *L. aspera* or with another species. Other hybrids are known in the genus, for example, between *L. spicata* and *L. aspera* and *L. cylindracea*, as has been demonstrated by Cruise (1964), Hadley and Levin (1967), and Levin (1967). Due to the similarity of morphology between these species, only detailed population studies are likely to confirm or refute these suggested or inferred hybridizations.

a.  Pappus barbellate, lateral cilia very short.
  b.  Rachis of inflorescence glabrous, or at most puberulent; heads sessile, spicate, phyllaries with scarious margins . . . . . . . . . . . . . . .1. *L. spicata*
  bb. Rachis of inflorescence pubescent (evident to the naked eye).
    c.  Inflorescence densely spicate, the distance between the heads rarely 1 cm; heads 4–18-flowered; phyllaries acuminate, the tips somewhat spreading, recurved . . . . . . . . . . . . . . . . . . . . . . . . .2. *L. pycnostachya*
    cc. Inflorescence more openly spicate to subracemiform, the heads short-pedunculate, 25–50-flowered; distances between heads 1–3 cm.
      d.  Phyllaries herbaceous, with narrow scarious margins, pubescent to rough on back, the outer squarrose; leaves and stem pubescent; basal leaves broadly obovate . . . . . . . . . . . . . . . . . . . . . . . . . . .3. *L. scariosa*

dd. Phyllaries thin, glabrous, broadly scarious, erose to lacerate, rough to glabrous; leaves and stems short-pubescent to glabrous; leaves linear to linear-lanceolate ................................4. *L. aspera*

aa. Pappus plumose.

b. Heads 3–8-flowered; corolla lobes glabrous within; leaves punctate .... ...........................................5. *L. punctata*

bb. Heads 15–60-flowered; corolla lobes hairy within.

c. Stems and peduncles pubescent; phyllaries lanceolate, pubescent, spreading, sharp-pointed, squarrose ...............6. *L. squarrosa*

cc. Stems and peduncles glabrous; phyllaries ovate, the outer appressed, glabrous on back, short-mucronate ...............7. *L. cylindracea*

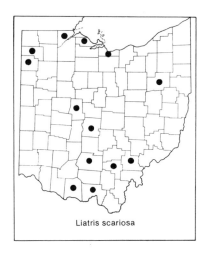

Liatris scariosa

### 1. **Liatris spicata** (L.) Willd.   Dense Blazing-star

Stems and leaves essentially glabrous; heads sessile and crowded along the elongated rachis; involucre subcylindric to turbinate-campanulate; pappus barbellate; corolla glabrous within, usually purplish (white-flowered forms have been reported). Wet prairie habitats. Midsummer–fall.

### 2. **Liatris pycnostachya** Michx.

Stems and leaves hirsute, especially along the elongated rachis; heads sessile, crowded, involucre subcylindric to turbinate-campanulate; corolla glabrous or essentially so within; pappus barbellate. Mostly on drier prairie sites. Midsummer–fall.

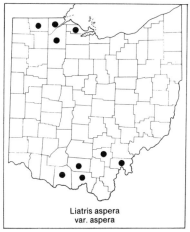

Liatris aspera
var. aspera

### 3. **Liatris scariosa** (L.) Willd.

Stems and leaves glabrous or at most slightly pubescent; heads on peduncles about 2–4 cm long; phyllaries broadly rounded, mostly with narrow, scarious, erose to lacerate margins; corolla pubescent within, especially toward base; pappus barbellate. Dry prairie sites. Aug.–fall.

Our most variable species of the genus, it can be separated from *L. aspera* only with some difficulty. The very narrow, hardly erose phyllary margins with rough to pubescent backs best characterize *L. scariosa*, while *L. aspera* has phyllaries that are thin, glabrous with broad scarious margins that are erose and more often lacerate. A few specimens from southern Ohio approach var. *nieuwlandii* Lunell.

### 4. **Liatris aspera** Michx.

Stems and leaves short-hairy or glabrous; leaves gradually reduced upward; inflorescence spiciform; involucre campanulate to subhemispheric, glabrous; phyllaries spreading, often a bit squarrose, purplish, erose, or often with lacerate scarious margins; corolla hairy within; pappus barbellate. Dry, open places, woodland borders, and prairie habitats. Common. Aug.–fall.

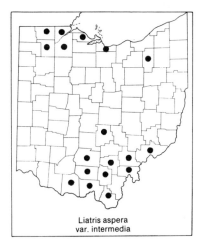

Liatris aspera
var. intermedia

x 1½

x 1½

x ½

Liatris spicata

x ½

Liatris pycnostachya

Liatris scariosa    × ½

Liatris aspera    × ½

Liatris squarrosa    × ½

x 1

x ½

Liatris cylindracea

Liatris squarrosa

Liatris cylindracea

*Liatris aspera* consists of two varieties in Ohio, var. *aspera* and var. *intermedia* (Lunell) Gaiser; var. *aspera* has appressed hairs on lower stems while var. *intermedia* is glabrous.

5. LIATRIS PUNCTATA Hook.    DOTTED BLAZING-STAR

This distinct species is included even though it is ordinarily more western in distribution. Long (1959) reports a single 1903 collection from Franklin County. I believe it to be merely a transient (not illustrated or mapped).

6. **Liatris squarrosa** (L.) Michx.    SCALY BLAZING-STAR

Heads pedunculate; stems and peduncles pubescent; phyllaries long, lanceolate, acuminate, squarrose; pappus plumose, visible even to the naked eye. Dry and wet prairie sites. July–early fall.

Of the three varieties listed by Cronquist (1952), only var. *squarrosa* is represented in the Ohio flora.

7. **Liatris cylindracea** Michx.

Easily distinguished from *L. squarrosa* by its glabrous stems and peduncles; phyllaries ovate, the outer abruptly sharp-pointed, appressed. Dry prairie habitats. July–early fall.


## TRIBE III. ASTEREAE CASS.

a.  Ray flowers yellow (white to cream-colored in two species of *Solidago*).
  b.  Pappus of disk flowers of 2–10 chaffy scales or short awns.
    c.  Heads 1 cm or more wide . . . . . . . . . . . . . . . . . . . . . . . . . .7. *Grindelia*
    cc. Heads small, less than 1 cm wide . . . . . . . . . . . . . . . . . .8. *Gutierrezia*
  bb. Pappus of disk flowers all or partly of capillary bristles.
    c.  Pappus of disk flowers double, the outer of short bristles, the inner of longer capillary bristles.
      d.  Lower leaves linear, veins appearing parallel; pubescence of stem and leaves long and silky; fruits linear . . . . . . . . . . . . . . . . . . .9. *Pityopsis*
      dd. Lower leaves broader, pinnately veined; pubescence various; fruits obconic.
        e.  Pubescence of lower stem and leaves of fine flagelliform hairs. . . . . .
        . . . . . . . . . . . . . . . . . . . . . . . . . . . . . . . . . . . . . . . .10. *Chrysopsis*
        ee. Pubescence of lower stem and leaves of rigid hairs . . . . . . . . . . . . . .
        . . . . . . . . . . . . . . . . . . . . . . . . . . . . . . . . . . . . . . .11. *Heterotheca*
    cc. Pappus of capillary bristles only . . . . . . . . . . . . . . . . . . . . . .12. *Solidago*
aa. Ray flowers white, pink, purple, or blue, never yellow.
  b.  Pappus absent or if present, of very short bristles and 2–4 longer awns.
    c.  Plants scapose, the leaves all basal . . . . . . . . . . . . . . . . . . . . .13. *Bellis*
    cc. Plants with leafy stems, not scapose. . . . . . . . . . . . . . . . . .14. *Boltonia*
  bb. Pappus of long capillary bristles.

Grindelia lanceolata

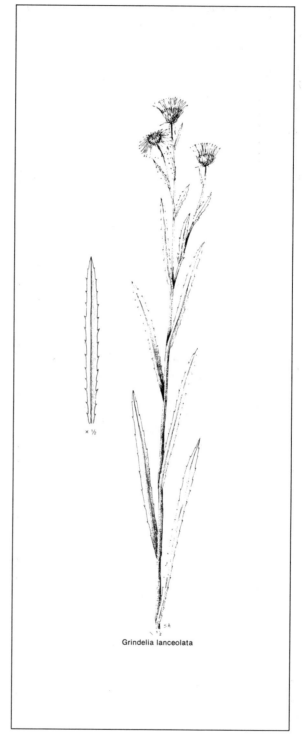

× ½

Grindelia lanceolata

c. Heads arranged on leafy branches; phyllaries strongly imbricated or with a few outer foliaceous series about as long as the inner.........
.............................................15. *Aster*

cc. Heads on essentially naked branches or peduncles; phyllaries in a single series or if in more than one series then the ray flowers scarcely longer than the disk.

d. Heads to about 5 mm wide; ray corollas about equaling or slightly longer than the disk corollas; disk flowers about 20 .....16. *Conyza*

dd. Heads more than 5 mm wide, usually exceeding 1 cm; ray corollas much longer than the disk corollas; disk flowers many, more than 20..........................................17. *Erigeron*

### 7. Grindelia Willd. TARWEED. GUMWEED

Heads (in ours) radiate. Ray flowers yellow, carpellate, fertile. Disk flowers yellow, often not seed-forming. Phyllaries imbricate or nearly equal, somewhat resinous. Receptacle convex or flat, naked. Pappus of 2-several deciduous awns or scales. Achenes compressed, slightly nerved, slightly scabrous or smooth. Biennial or perennial herbs often woody at caudex. Leaves alternate, resin-dotted, serrate to nearly entire.

Our species are introduced from western and southwestern North America.

a. Phyllary tips erect, scarcely resinous; leaf margins entire or serrate with bristle-tipped teeth..............................1. *G. lanceolata*

aa. Phyllary tips reflexed, abundantly resinous; leaf margins entire or with blunt or rounded teeth..........................2. *G. squarrosa*

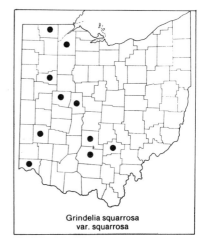

Grindelia squarrosa
var. squarrosa

1. GRINDELIA LANCEOLATA Nutt.
Stems glabrous to slightly hairy, corymbosely branched; leaves scarcely punctate, entire or serrate with bristle-tipped teeth, linear; heads several; phyllaries slightly viscid, loose, ascending, not squarrose; achenes smooth, often lustrous; pappus of 2 awns.

Known from a single specimen in The Ohio State University Herbarium collected on Vore Ridge in Athens County. Aug.–Sept.

2. GRINDELIA SQUARROSA (Pursh) Dunal
Stems stout, corymbosely branched: leaves strongly punctate, crenate to serrate or entire, linear to ovate or narrowly obovate; heads large, to 2 cm in diameter; phyllaries glutinous, tips strongly recurved; achenes striate; pappus 2–8. Introduced from western United States and adventive in our area, where it grows in dry habitats along roadsides and in waste places. July–Sept.

Two varieties can be recognized.

a. Upper and middle leaves oblong to narrow-obovate; rays 1 cm or less ...
..............................................var. *squarrosa*

aa. Upper and middle leaves lanceolate to oblanceolate; rays 1 cm or longer (illustrated)........................var. *serrulata* (Rydb.) Steyerm.

Grindelia squarrosa
var. serrulata

Grindelia squarrosa

### 8. Gutierrezia Lag. BROOM-SNAKEROOT

Heads numerous in a flat-topped arrangement, radiate, carpellate and seed-forming, yellow. Phyllaries imbricate. Receptacle naked, pitted. Disk flowers perfect, apparently sterile. Achenes striate, roundish. Pappus a minute crown in ray flowers, composed of awns or narrow scales in disk flowers. Taprooted annual, glabrous or often glutinous. Leaves alternate, linear.

1. GUTIERREZIA DRACUNCULOIDES (DC.) Blake
Represented by a single collection from Seneca County where it grows in a rather dry habitat (not illustrated or mapped). Midsummer–fall.

## 9. Pityopsis Nutt.

Heads radiate. Receptacle flat or slightly convex, naked. Disk flowers perfect, fertile. Ray flowers yellow, carpellate, seed-forming. Pappus double, the inner of capillary bristles, the outer of short bristles or scales. Achenes linear to fusiform. Fibrous-rooted perennials often with stoloniferous rhizomes. Stems and leaves with long, silky and often silvery hairs. Native to North America, distributed primarily in the southeastern states, with one species reaching Ohio.

Pityopsis graminifolia

Pityopsis graminifolia

1. **Pityopsis graminifolia** (Michx.) Nutt.   SILKGRASS
*Chrysopsis graminifolia* (Michx.) Ell.
*Chrysopsis microcephala* Small
*Heterotheca graminifolia* (Michx.) Shinners
*Heterotheca microcephala* (Small) Shinners
*Pityopsis microcephala* (Small) Small

This species is often confused with *Chrysopsis* or *Heterotheca*, but is separable from them by the presence of long, silky hairs and fusiform fruits. Known primarily from the southern counties of Adams and Scioto, where it occurs in dry, sandy habitats. July–Sept.

Traditionally, the species of *Pityopsis*, *Chrysopsis*, and *Heterotheca* that occur in Ohio have all been placed within *Chrysopsis sensu lato*. The present treatment of these genera follows the work of Semple, Blok, and Heiman (1980) and, in part, Semple (1977).

## 10. Chrysopsis (Nutt.) Ell.   GOLDEN ASTER

Heads radiate. Phyllaries imbricate. Receptacle flat or slightly convex, naked. Disk flowers perfect, seed-forming, yellow. Ray flowers yellow, carpellate, seed-forming. Pappus double, the inner of capillary bristles, the outer of scales or bristles. Achenes flattened, obconic, somewhat pubescent. Perennial, silky-pubescent herbs. Leaves alternate, sessile. Native to North America, with one species occurring in Ohio.

1. **Chrysopsis mariana** (L.) Ell.
*Heterotheca mariana* (L.) Shinners

Stems and herbage with fine flagelliform hairs, often becoming glabrate with age, plants usually with several stems from a thick base; lower leaves oblanceolate to obovate, petiolate, the middle and upper gradually reduced, lanceolate and sessile; heads in a corymbose arrangement, often crowded; peduncles glandular; phyllaries glandular with ascending tips, white-hairy; achenes obovate, slightly hairy. Grows in acid sand or gravelly soils. Aug.–Oct.

For a study of cytology, see Semple and Chinnappa (1980).

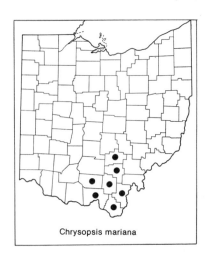

Chrysopsis mariana

## 11. Heterotheca Cass.   GOLDEN ASTER

Heads radiate, corymbiform. Phyllaries imbricate, tapering to minute tips. Receptacle flat or slightly convex. Ray flowers yellow. Disk flowers perfect and seed-forming. Achenes obconic, faintly nerved. Pappus double, the outer longer. Mostly tap-rooted perennials. Stems and leaves with rigid osteolate hairs. Leaves chiefly cauline, elliptic to oblong, sessile, the lower mostly oblanceolate to obovate, usually with a few small teeth. Native to the United States and Mexico. See Harms (1974) for a preliminary conspectus of the genus.

1. Heterotheca camporum (Greene) Shinners

    *Chrysopsis camporum* Greene

    *Chrysopsis villosa* (Pursh) Nutt. var. *camporum* (Greene) Cronq.

    Dry waste ground. July–Oct.

    Probably introduced from Missouri, Illinois, and Indiana.

× 1½

× 1

x ½

Chrysopsis mariana

×5    ×1

×1

Heterotheca camporum

× ½

Heterotheca camporum

Heterotheca camporum

## 12. Solidago L.  GOLDENROD

Heads radiate, campanulate to subcylindric, the ray flowers ovulate and seed-forming, yellow (white or cream-colored in two species), disk flowers perfect, seed-forming. Receptacle naked, flat or slightly convex. Achenes several-nerved, slightly flattened, glabrous or pubescent. Pappus double, of capillary bristles. Leaves alternate, toothed or entire, few to numerous.

The recognition of the genus *Euthamia* Nutt., often treated as a section of *Solidago*, is gaining in popularity. *Euthamia* includes those species with flat-topped (corymbose) inflorescences (Sieren, 1981). This segregation may well be justified because those species, according to Cronquist (1980), appear more closely related to other genera than to *Solidago*. However, I have elected to retain this group within the genus *Solidago*, pending the outcome of further taxonomic investigations.

There are reports of *Solidago arguta* Ait. and *S. jacksonii* (Ktze.) Fern. from Ohio, but I have seen no specimens.

a.  Heads in corymbose arrangement.
b.  Ray flowers white . . . . . . . . . . . . . . . . . . . . . . . . . . . . .1. *S. ptarmicoides*
bb. Ray flowers yellow.
   c.  Upper leaves narrow, rarely 5 mm wide, lower leaves deciduous by anthesis.
      d.  Leaves 3-nerved; upper stems nearly always pubescent; heads sessile
      . . . . . . . . . . . . . . . . . . . . . . . . . . . . . . . . . . . . . . . . . .2. *S. graminifolia*
      dd. Leaves not 3-nerved or if so, the two lateral nerves very faint; upper stems glabrous; heads pedunculate to subsessile . . . . . . . . . . . . . . . .
      . . . . . . . . . . . . . . . . . . . . . . . . . . . . . . . . . . . . .3. *S. gymnospermoides*
   cc. Upper leaves wider than 5 mm, lower leaves usually present at anthesis.
      d.  Leaves ovate-oblong to elliptic-oblong, gradually reduced above, densely short-pubescent, grayish in appearance. . . . . . . . . .4. *S. rigida*
      dd. Leaves linear to lanceolate.
         e.  Leaves oblong-lanceolate, much reduced upward, toothed or entire, flat, not 3-nerved . . . . . . . . . . . . . . . . . . . . . . . . . . .5. *S. ohioensis*
         ee. Leaves linear-lanceolate, folded, not conspicuously reduced upward, 3-nerved or at least faintly so. . . . . . . . . . . . . . . . . . . . .6. *S. riddellii*
aa. Heads in a racemose, thyrsoid, or paniculate arrangement.
b.  Phyllaries, or at least the outer, distinctly squarrose. . . . .7. *S. squarrosa*
bb. Phyllaries appressed, erect.
   c.  Inflorescences axillary or terminal, the heads in clusters or short racemes from axils of rather well-developed leaves or large bracts.
      d.  Stems densely pubescent (the pubescence visible to the naked eye).
         e.  Rays white to cream. . . . . . . . . . . . . . . . . . . . . . . . . . . .8. *S. bicolor*
         ee. Rays yellow. . . . . . . . . . . . . . . . . . . . . . . . . . . . . . . . .9. *S. hispida*
      dd. Stems glabrous.
         e.  Leaves ovate, sharply serrate; stems usually angled and zigzag . . . . . .
         . . . . . . . . . . . . . . . . . . . . . . . . . . . . . . . . . . . . . . . .10. *S. flexicaulis*

ee. Leaves lanceolate, serrate, tapering to base, sessile or subsessile; stems glaucous, not angled . . . . . . . . . . . . . . . . . . . . . . . . . .11. *S. caesia*

cc. Inflorescences not axillary, either secund, paniculate, racemose, or spicate.

  d.  Inflorescences dense, either narrowly paniculate or narrowly racemose, often interrupted; plants of dry habitats.

    e.  Lowest cauline leaves widest at about the middle or slightly below, usually well over 2 cm wide; inflorescence branches usually stiff and ascending . . . . . . . . . . . . . . . . . . . . . . . . . . . . . . . . .12. *S. speciosa*

    ee. Lowest cauline leaves widest slightly above the middle, usually less than 2 cm wide; inflorescence branches somewhat spreading . . . . . . . . . . . . . . . . . . . . . . . . . . . . . . . . . . . . . . . . . . . . . . . . . . . . . . . . . . . . . . .13. *S. erecta*

dd. Inflorescences distinctly secund, i.e., one-sided, racemes.

  e.  Leaves 1- or more often 3-nerved.

    f.  Stems glabrous except in the inflorescence.

      g.  Stems sharply angled; leaves scabrous above with stiff, pustular-based hairs. . . . . . . . . . . . . . . . . . . . . . . . . . . . . . . . .14. *S. patula*

      gg. Stems terete, not sharply angled. . . . . . . . . . . . . . . .15. *S. gigantea*

    ff. Stems pubescent in and below the inflorescence.

      g.  Leaves oblanceolate; basal rosettes present at anthesis; stems with loose, spreading pubescence . . . . . . . . . . . . . . . . .16. *S. nemoralis*

      gg. Leaves lanceolate, acuminate; basal rosettes absent at anthesis; stems scabrous to minutely villous with fine, gray pubescence . . . . . . . . . . . . . . . . . . . . . . . . . . . . . . . . . . . . . . . . . . . .17. *S. canadensis*

ee. Leaves with more than 3 nerves, usually pinnately veined.

  f.  Stems pubescent.

    g.  Leaves oblanceolate, crenate-dentate or entire, rapidly diminishing in size upward; rosette leaves present at anthesis; short, leafy axillary branches often present; stems often gray-pubescent . . . . . . . . . . . . . . . . . . . . . . . . . . . . . . . . . . . . . . . . . . . . . . . . . .16. *S. nemoralis*

    gg. Leaves elliptic-lanceolate or broader, tapering at both ends, not rapidly decreasing in size upward; basal leaves soon deciduous . . . . . . . . . . . . . . . . . . . . . . . . . . . . . . . . . . . . . . .18. *S. rugosa*

  ff. Stems glabrous, or at most puberulent in lines.

    g.  Leaves pellucid-punctate, anise-scented when crushed, entire, sessile; stems puberulent in lines or nearly glabrous . . . .19. *S. odora*

    gg. Leaves not punctate, not anise-scented, at least some leaves toothed, sessile or petiolate; stems glabrous below inflorescence.

      h.  Leaves clasping or sheathing at base; plants of bogs and wet places . . . . . . . . . . . . . . . . . . . . . . . . . . . . . . . . .20. *S. uliginosa*

      hh. Leaves not clasping or sheathing.

i. Leaves scabrous above with pustular-based hairs; stems angled; achenes pubescent ..........................14. *S. patula*

ii. Leaves glabrous or pubescent above, but not scabrous with pustular hairs; stems terete, not angled; achenes pubescent or glabrous.

  j. Leaves pubescent on upper surface with scattered hairs, more densely so beneath, leaves often reduced upward, basal leaves absent at anthesis......................21. *S. ulmifolia*

  jj. Leaves glabrous or only slightly pubescent, basal leaves present at anthesis, long-acuminate, tapering to margined petioles, the upper leaves often with tufts in the axils........22. *S. juncea*

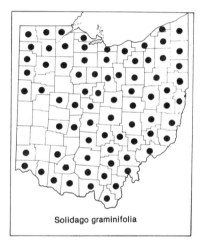

Solidago ptarmicoides

1. **Solidago ptarmicoides** (Nees) Boivin

  *Aster ptarmicoides* (Nees) T. & G.

Stems to 1 m tall, scabrous above; leaves entire or with a few scattered teeth, often faintly 3-nerved, the lower leaves petiolate, linear-oblanceolate and much larger than the upper, the upper becoming sessile; heads arranged in an open corymb, often bracteate; phyllaries imbricated, midribs strong; ray flowers numbering to 25, white; disk flowers white; pappus bristles white, clavate at summit. A rare species of sandy habitats, known from only three northwestern counties (not illustrated). July–Sept.

According to Brouillet and Semple (1981) an abundance of hybrids exists between this taxon and other species of *Solidago*, especially species of section *Oligoneuron*. Those authors suggest, therefore, that this species be removed from the genus *Aster*, where it was formerly placed, and be included within the genus *Solidago*.

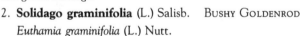

Solidago graminifolia

2. **Solidago graminifolia** (L.) Salisb.  Bushy Goldenrod

  *Euthamia graminifolia* (L.) Nutt.

Stems usually pubescent; rhizomes branched; leaves linear to lanceolate, acuminate, glandular-punctate, 3-nerved; inflorescence terminal, corymbose, the heads sessile in glomerules; phyllaries with green tips; achenes with short hairs. Widely distributed in Ohio, where it occurs in open, usually moist habitats. July–Oct.

Two weakly separated varieties may be recognized in plants of our area: var. *nuttallii* Greene and var. *graminifolia*. Of the two, var. *nuttallii* (illustrated), distinguished by its denser stem pubescence, is the more common.

3. **Solidago gymnospermoides** (Greene) Fernald

  *Euthamia gymnospermoides* Greene

  *Solidago moseleyi* Fernald

  *Solidago remota* (Greene) Friesner

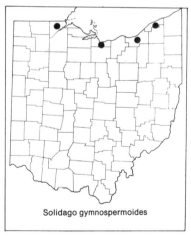

Solidago gymnospermoides

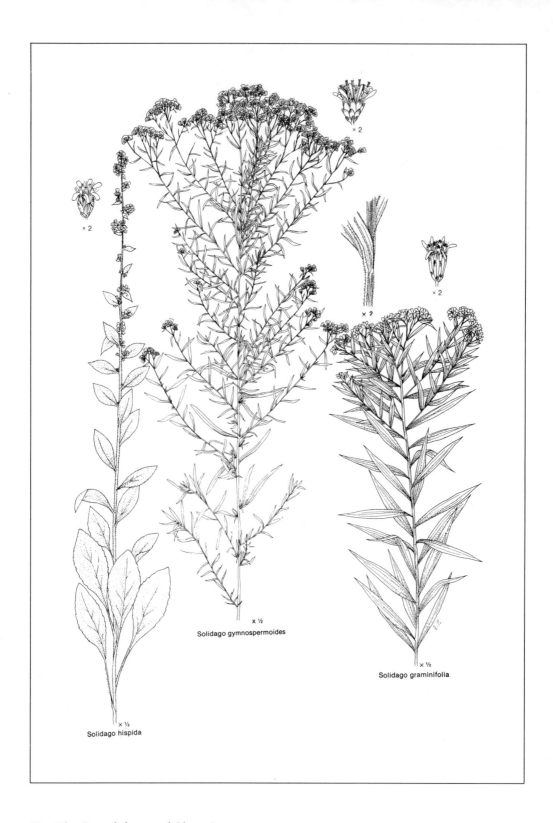

× 2

× 2

× 2

× 2

x ½
Solidago gymnospermoides

x ½
Solidago graminifolia

x ½
Solidago hispida

Stems glabrous from creeping rhizomes; leaf surfaces glabrous, glandular-punctate, faintly 3-nerved; inflorescence terminal, heads arranged in a corymb, usually pedunculate to subsessile in small glomerules; achenes with short hairs. Distribution restricted to the northern counties of Ohio, where it occurs mostly on sandy soils near Lake Erie and in the Oak Openings of Lucas County. Mid-Aug.–Sept.

This species is similar to *S. graminifolia*, but differs in its glabrous stems and only faintly 3-nerved leaves. For a study of the *S. graminifolia–S. gymnospermoides* complex, see Croat (1970).

### 4. **Solidago rigida** L.  STIFF GOLDENROD

Stems stout, stiff, hairy, the dense, short, spreading hairs presenting a gray appearance; leaves elliptic to elliptic-oblong or ovate-oblong, nearly entire or slightly toothed, gradually reduced upward to the inflorescence; inflorescence terminal, heads arranged in a corymb and larger than those of most goldenrods; achenes angled, glabrous or only slightly pubescent. Sporadically distributed in Ohio, in prairies and open habitats, especially in sandy areas. Aug.–Oct.

Only var. *ridiga* has been observed in Ohio.

### 5. **Solidago ohioensis** Riddell  OHIO GOLDENROD

Stems glabrous; leaves glabrous except for ciliate margins, basal and cauline leaves persistent and well-developed, mostly oblong-lanceolate, progressively reduced upward and into the inflorescence; lower and basal leaves with long petioles, but petioles shorter upward; inflorescence terminal, corymbose; phyllaries glabrous, often rounded; achenes angled, glabrous. Common in moist habitats, especially swamps, bogs and wet prairies, and habitually wet ditches. Aug.–Sept.

This species is easily distinguished from the other species of the genus by its persistent basal leaves and progressively reduced lanceolate to linear and glabrous upper leaves. See Pringle (1982) for a study of the species' distribution.

### 6. **Solidago riddellii** Frank  RIDDELL'S GOLDENROD

Stems glabrous, at least below the inflorescence; leaves glabrous except for the ciliate margins, entire, conduplicate, faintly 3-nerved, the lower petiolate, the middle and upper sessile, clasping at base and not generally reduced upward; inflorescence terminal, corymbose; phyllaries glabrous, obtuse or broadly rounded; achenes slightly angled, glabrous. Common in moist habitats, especially swamps, bogs, and wet prairies. Aug.–Sept.

The folded and somewhat clasping, glabrous leaves are good field characters.

### 7. **Solidago squarrosa** Muhl.  STOUT GOLDENROD

Stems glabrous, at least below the inflorescence; leaves essentially glabrous to

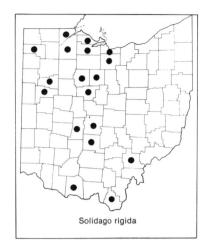

Solidago rigida

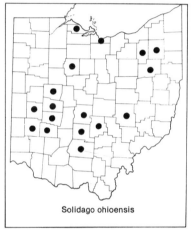

Solidago ohioensis

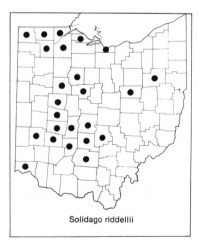

Solidago riddellii

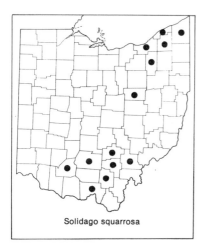

Solidago squarrosa

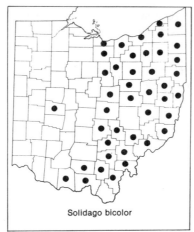

Solidago bicolor

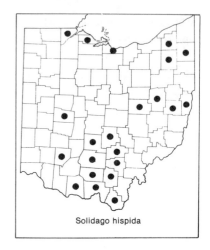

Solidago hispida

somewhat scabrous above, basal leaves persistent, all leaves oblanceolate to obovate, serrate, tapering onto petiole, cauline leaves reduced upward; inflorescence narrow, commonly leafy-bracteate; outer phyllaries squarrose; achenes glabrous. A plant of woodlands, primarily and easily distinguished by the spreading and recurved tips of the phyllaries. Aug.–early Oct.

8. **Solidago bicolor** L.   White Goldenrod. Silver-rod

Stems stout, pubescent with spreading-hirsute hairs; leaves well-developed and persistent, broadly oblanceolate to elliptic, serrate (sometimes merely crenate), tapering onto petiole at base, reduced and sessile upwards; inflorescence terminal, the branches narrow and ascending; phyllaries light colored; ray flowers white or creamy, sometimes appearing yellowish in dried specimens; achenes glabrous. Common in open, dry woodlands. July–Oct.

The pubescent stems with blades tapering onto the petioles and the white ray flowers are the distinguishing characters.

9. **Solidago hispida** Muhl.   Hairy Goldenrod

Stems spreading-hirsute; basal and cauline leaves persistent, broadly oblanceolate to obovate, serrate, sometimes merely crenate or entire, petiolate, the blade tapering onto the petiole, the upper leaves reduced and becoming sessile; inflorescence terminal, elongate and narrow, somewhat leafy-bracteate, stiffly ascending; ray flowers rich yellow; achenes glabrous. Common in dry, upland woods. July–Oct.

Except for the color of the rays, similar to S. *bicolor.*

10. **Solidago flexicaulis** L.   Zigzag Goldenrod

Stems from creeping rhizomes, stout, angled and grooved, and zigzag, especially in the upper stem, glabrous, at least below the inflorescences; leaves very sharply serrate, hirsute beneath, seldom glabrous, glabrous or with a few hairs above; lower and basal leaves deciduous at anthesis, middle cauline

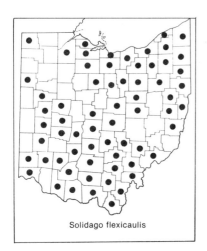

Solidago flexicaulis

× 1

× ½
Solidago rigida

× 2

× ½
Solidago ohioensis

× 2

× ½
Solidago riddellii

× 2

× 2

× ½
Solidago squarrosa

× ½
Solidago bicolor

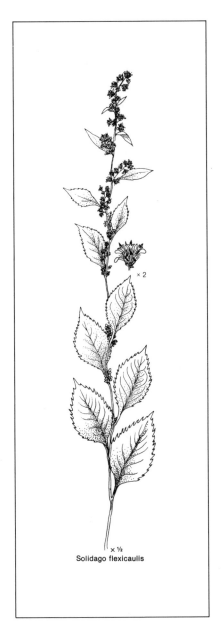

× 2

× ½
Solidago flexicaulis

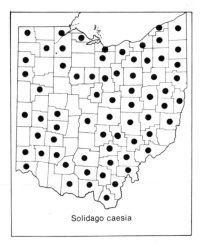

Solidago caesia

leaves ovate to elliptic, petiole winged; inflorescences narrow, racemose, heads somewhat clustered in axils of barely reduced leaves; phyllaries glabrous except for ciliate margins; achenes with short hairs. Common throughout Ohio in woodland habitats. July–Oct.

11. **Solidago caesia** L.   BLUE-STEM GOLDENROD
Stems glabrous and glaucous; leaves serrate, essentially glabrous, lower leaves

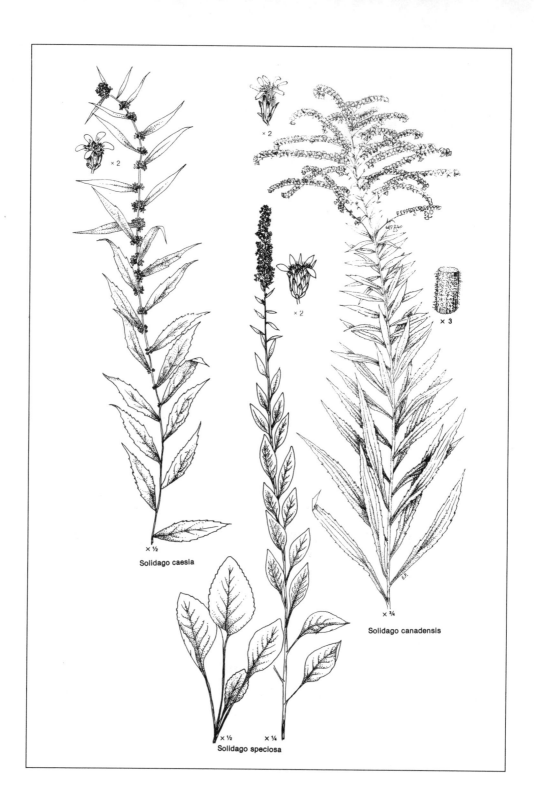

× 2

× 2

× 2

× 3

× ½

Solidago caesia

× ¾

Solidago canadensis

× ½         × ¼

Solidago speciosa

deciduous at anthesis, middle cauline leaves lanceolate, tapering to sessile base, upper leaves slightly reduced; inflorescences in axils of reduced leaves, narrow; phyllaries glabrous; ray flowers few, 4 or 5; achenes pubescent. Throughout Ohio, especially common in woodland habitats. Aug.–Oct.

## 12. **Solidago speciosa** Nutt.   SHOWY GOLDENROD

Stems stout, glabrous except in the inflorescence; lower leaves persistent and large, middle cauline leaves glabrous, mostly entire, sometimes crenate, ovate-lanceolate, decurrent onto petiole; inflorescence a dense, pyramidal to thyrsiform panicle; phyllaries firm, glabrous; achenes glabrous. Infrequent, in open fields and prairie-like habitats. Late Aug.–Oct.

This species is generally separated into three varieties, two of which occur in Ohio. Variety *rigidiuscula* T. & G. is well represented in Lucas County in the Oak Openings, but rarely found elsewhere. The robust var. *speciosa* is the more frequent, but found in only a few counties. The former, var. *rigidiuscula*, is quite distinct, with extremely reduced upper leaves and slightly scabrous stems.

## 13. **Solidago erecta** Pursh   SLENDER GOLDENROD

Stems glabrous except in the inflorescence; lower leaves oblong-ovate, persistent at anthesis, middle leaves oblanceolate to obovate, mostly crenate, sometimes nearly entire, becoming reduced and sessile upwards; inflorescence narrow, elongate; phyllaries rounded, glabrous; achenes glabrous. Dry or moist woods and thickets, primarily in southeastern Ohio. Aug.–Oct.

Plants of this species may be confused with *S. hispida*, but differ in having glabrous stems.

## 14. **Solidago patula** Muhl.   ROUGH-LEAVED GOLDENROD

Stems glabrous below the inflorescence, angled; leaves scabrous above with pustular-based hairs, glabrous beneath; lower leaves large, persistent, elliptic-ovate, serrate; middle leaves ovate-lanceolate, serrate; upper leaves reduced; inflorescence terminal with spreading, recurved, secund branches; phyllaries ciliate, green; achenes minutely hairy. Common in bogs, swamps, wet meadows, and on flood plains. Aug.–Sept.

Ohio plants belong to var. *patula*.

## 15. **Solidago gigantea** Ait.

Stems stout, glabrous below inflorescence, sometimes glaucous; leaves 3-nerved, mostly glabrous above, glabrous or pubescent on main veins beneath; lower leaves deciduous; middle leaves rather uniform in size; upper leaves lanceolate, only slightly reduced; inflorescence terminal, paniculate, secund; phyllaries blunt, often green-tipped; achenes with short pubescence. Common in moist, open habitats. July–Sept.

Two varieties may be recognized: var. *gigantea* has the midrib and lateral

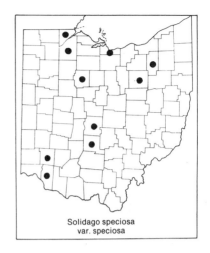

Solidago speciosa
var. speciosa

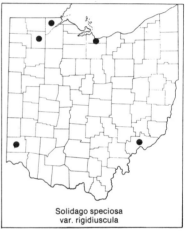

Solidago speciosa
var. rigidiuscula

Solidago erecta

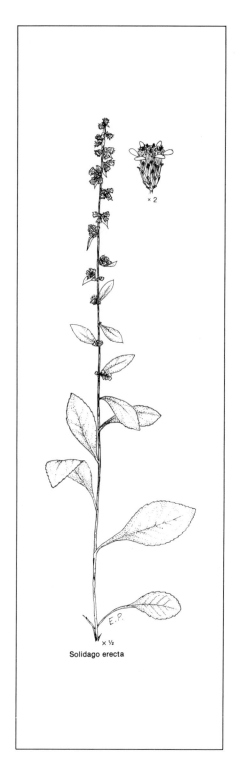

× 2

× ½

**Solidago erecta**

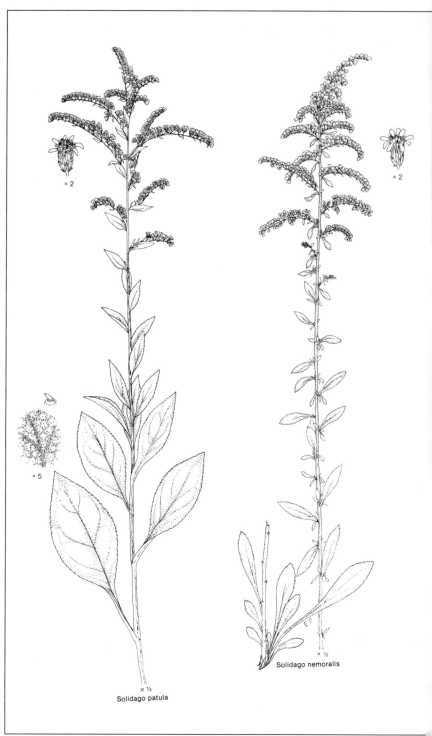

× 2

× 5

× 2

× ½

**Solidago patula**

× ½

**Solidago nemoralis**

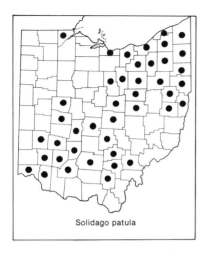

Solidago patula

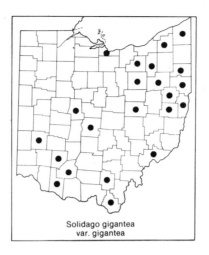

Solidago gigantea
var. gigantea

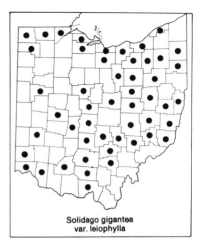

Solidago gigantea
var. leiophylla

veins pubescent on lower leaf surface, while var. *leiophylla* Fern. (illustrated) is glabrous on both surfaces. The species appears to intergrade with *S. canadensis* L.

### 16. **Solidago nemoralis** Ait.  GRAY GOLDENROD

Stems gray-pubescent with fine and loosely spreading hairs; axillary tufts sometimes present; leaves puberulent on both surfaces; lower leaves well-developed, long-petiolate, crenate-dentate to entire, often deciduous; leaves gradually reduced and becoming sessile upwards; inflorescence paniculate with divergent, recurved, secund branches; phyllaries ciliate, otherwise glabrous; achenes pubescent. Common in dry, open woodlands. Early Aug.–Oct.

Although somewhat variable, our plants of this species are assignable to var. *nemoralis*.

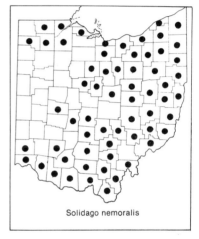

Solidago nemoralis

### 17. **Solidago canadensis** L.  CANADA GOLDENROD

incl. *S. altissima* L.

Stems tall, pubescent below the inflorescence; leaves 3-nerved, lanceolate, sharply serrate to nearly entire, acuminate, glabrous to slightly scabrous above, glabrous to more commonly puberulent on midrib and main veins beneath, with sparing, scattered pubescence elsewhere; inflorescence paniculate with conspicuously recurved, secund branches; phyllaries linear, yellowish-green; achenes short-pubescent. Common in a variety of habitats, such as open places, cut-over areas, thickets, roadsides, and fields. July–Sept.

In this treatment I have elected to include *S. altissima* within *S. canadensis*, since no combination of characters seems to separate the two satisfactorily, at least among Ohio plants. In a few locations in southern Ohio it is possible to separate var. *rupestris* (Raf.) Cronq., with leaves consistently glabrous beneath and heads with 11 or fewer rays.

Solidago canadensis

×2

×2

×2

× ½
Solidago gigantea

× ½
Solidago juncea

× ½
Solidago ulmifolia

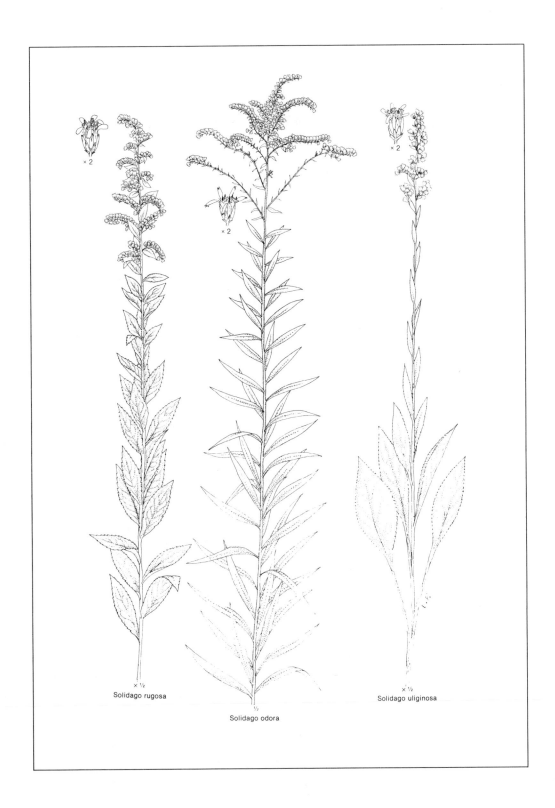

× 2

× 2

Solidago rugosa

× ½

Solidago odora

½

× 2

× ½

Solidago uliginosa

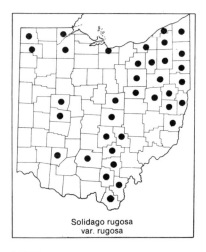

Solidago rugosa
var. rugosa

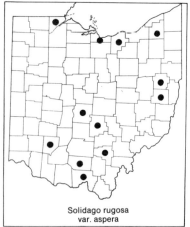

Solidago rugosa
var. aspera

Solidago odora

### 18. **Solidago rugosa** Mill.  Wrinkled-leaf Goldenrod

Stems pubescent with spreading hairs; lower leaves deciduous, not present at anthesis, only slightly reduced upwards, usually strongly rugose-veiny beneath, lanceolate, serrate, sessile to subsessile, tapering to base; inflorescence paniculate with recurved, secund branches; phyllaries glabrous; achenes pubescent. A variable species common in a variety of habitats but primarily in moist places. Aug.–Oct.

Two varieties may be distinguished, var. *rugosa* (illustrated) by its weakly rugose leaves with long, soft pubescence, and var. *aspera* (Ait.) Fern. by its strongly rugose leaves and short, harsh pubescence.

### 19. **Solidago odora** Ait.  Fragrant Goldenrod

Stems puberulent in lines from leaf bases; leaves glabrous except along margins, pellucid-punctate, anise-scented when crushed, basal leaves reduced and deciduous, middle leaves lanceolate, upper leaves somewhat reduced, only midrib prominent beneath; inflorescence paniculate with widely divergent and recurving, secund branches; phyllaries unequal, yellowish; achenes short-hairy. Found only in a few southern counties, where it occurs in dry, open woodlands. July–Sept.

### 20. **Solidago uliginosa** Nutt.  Bog Goldenrod

Stems glabrous except in the inflorescence; basal leaves persistent at anthesis, oblanceolate, subentire to serrate, tapering to long petiole, somewhat sheathing at base, middle leaves lance-ovate, entire, glabrous, upper leaves reduced; inflorescence narrowly paniculate, the branches usually recurved, secund; phyllaries unequal, rounded or blunt, glabrous; achenes glabrous. The plants are somewhat variable and found mostly in acid or calcareous bogs, swamps, and wet meadows. July–Sept.

### 21. **Solidago ulmifolia** Muhl.  Elm-leaved Goldenrod

Stems essentially glabrous below inflorescence; leaves sharply and unevenly serrate, hirsute above, sparsely hirsute on veins and surface beneath, lower leaves deciduous at anthesis, middle leaves ovate, tapering at base, sessile or subsessile, upper leaves reduced and less serrate; inflorescence paniculate, the branches long and recurving-secund; achenes minutely pubescent. Common in woods and thickets. Aug.–Oct.

### 22. **Solidago juncea** Ait.  Plume Goldenrod

Stems glabrous; basal leaves persistent, tufted, linear-lanceolate, serrate, tapering to long petiole, cauline leaves lanceolate, becoming sessile, narrower upwards, glabrous except for ciliate margins, rarely somewhat scabrous; inflorescence paniculate with recurved-secund branches; phyllaries unequal, ciliate, often glutinous; achenes pubescent or glabrous. Common in open woods and dry habitats. June–Oct.

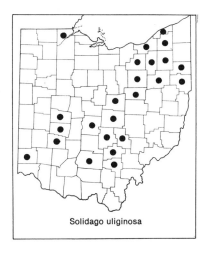

Solidago uliginosa

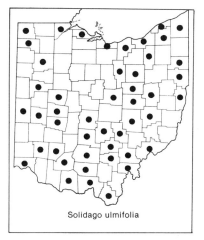

Solidago ulmifolia

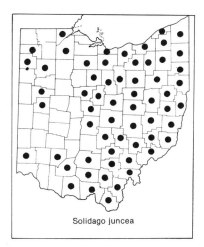

Solidago juncea

### 13. Bellis L. EUROPEAN DAISY. ENGLISH DAISY

Plants small, scapose; heads solitary, radiate. Phyllaries equal, in 1 or 2 series, herbaceous. Receptacle conic, naked. Ray flowers carpellate, white or pink. Disk flowers perfect, yellow. Achenes compressed. Pappus capillary or wanting. A small genus of Europe.

1. BELLIS PERENNIS L.

This attractive perennial species is often a weed in lawns or waste places and apparently naturalized in the northern portions of the United States. A few cultivars have been developed and are often used in low garden borders. April–Nov.

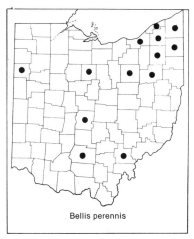

Bellis perennis

### 14. Boltonia L'Her. BOLTONIA

Heads radiate, ray flowers many, white, blue, or pink, carpellate and seed-forming. Phyllaries imbricate, acute or spatulate and rounded, midrib with broad green stripe. Receptacle conic or slightly convex, naked. Disk flowers perfect, seed-forming, yellow. Achenes obovate, flat, winged. Pappus of short scales or bristles and 2–4 long awns. Leaves variable but usually narrow and entire.

1. **Boltonia asteroides** L'Her.

Along streams and marshes primarily in northwestern counties. Aug.–Sept.

This species is represented in Ohio by var. *recognita* (Fern. & Grisc.) Cronq. Morgan (1967) reports possible hybrids between var. *recognita* and var. *asteroides* in Putnam County, but I have not seen these specimens. Plants of var. *recognita* usually have more than 25 heads, while those of var. *asteroides* usually have fewer than 25.

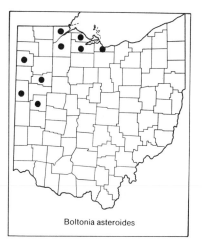

Boltonia asteroides

Tribe III. Astereae 65

× 1

Bellis perennis

1

× 5

× ½

Boltonia asteroides

## 15. Aster L.    ASTER

Heads numerous, hemispheric to subhemispheric, radiate. Phyllaries in two or more series, somewhat imbricate and of different lengths. Receptacle flat or slightly convex, naked. Disk flowers perfect, seed-forming, corollas usually yellow, sometimes brownish-purple, seed-forming. Ray flowers carpellate and seed-forming, corollas white, pink, purple, or blue. Pappus of capillary bristles, sometimes with an outer series present and shorter. Perennial herbs (annual in *A. subulatus*). Leaves alternate, entire or variously toothed.

There are probably 200 or more species of *Aster*, primarily in North America, but also extending into South America and rather widely distributed in the Old World. For a complete classification of the New World species, see Jones (1980); for a monograph of the genus in Ohio, see Speer (1958). For nomenclatural changes, see Jones (1983). I am treating Aster in the broad sense favored by Jones and Young (1983), who argue against the segregation of *Virgulus* Raf. (=*Lasallea* Greene, see Semple and Brouillet, 1980).

I have seen single specimens of *A. tradescanti* L. from Butler County and *A. acuminatus* Michx. from Ashtabula County, but I believe that these were waifs and that these species are not presently represented in the Ohio flora. Accordingly, they are not included in this treatment. *Aster tataricus* L.f. is often cultivated and escapes, but apparently does not persist in nature.

a.  Basal and lower cauline leaves cordate or subcordate at base *and* also petiolate.

  b.  Heads arranged in a corymb; plants rhizomatous, often forming large colonies; leaves usually serrate with mucronate teeth.

    c.  Phyllaries and peduncles with glandular hairs; phyllaries broad, mostly more than 1.5 mm wide; ray flowers normally violet to purplish . . . . . . . . . . . . . . . . . . . . . . . . . . . . . . . . . . . . . . . . . . . . . . . . . .1. *A. macrophyllus*

    cc. Phyllaries and peduncles eglandular although pubescence often present in inflorescence; phyllaries often less than 1.5 mm wide; ray flowers normally white, but sometimes becoming violet with age.

      d.  Lower leaves cordate, sinus of lower leaves rectangular; phyllaries few, narrow, about 1 mm wide; stems not zigzag . . . . . . . . .2. *A. schreberi*

      dd. Lower leaves cordate, sinus of lower leaves rounded; phyllaries to 1.5 mm wide; stems often zigzag . . . . . . . . . . . . . . . . . .3. *A. divaricatus*

  bb. Heads arranged in an elongate panicle; rays blue, violet, or lilac.

    c.  Cauline leaves sessile and cordate-clasping, *or* if petiolate, with clasping petioles; tips of phyllaries rhombic . . . . . . . . . . . . . . .4. *A. undulatus*

cc. Cauline leaves not clasping at base.

  d.  Leaves entire or nearly so.

    e.  Phyllaries pubescent on back; leaves nearly glabrous, only slightly scabrous above; nearly all leaves cordate . . . . . . . . . . . .5. *A. shortii*

    ee.  Phyllaries glabrous on back; leaves scabrous above, less so beneath; only the lower leaves cordate or subcordate . . . .6. *A. oolentangiensis*

  dd.  Leaves, at least the lower ones, serrate or dentate.

    e.  Herbage essentially glabrous throughout except for some puberulence in the inflorescence.

      f.  Phyllaries linear-lanceolate, attenuate, tips with narrow green stripe . . . . . . . . . . . . . . . . . . . . . . . . . . . . . . . . . . . . . .7. *A. sagittifolius*

      ff.  Phyllaries linear, obtuse, tips with broad diamond-shaped green area . . . . . . . . . . . . . . . . . . . . . . . . . . . . . . . .10. *A. lowrieanus*

    ee.  Herbage hairy, leaves scabrous or hairy.

      f.  Leaves sharply toothed, petioles only obscurely winged; phyllaries obtuse, often purple-tipped . . . . . . . . . . . . . . . . . . . .9. *A. cordifolius*

      ff.  Leaves broadly toothed, petioles obviously winged; phyllaries attenuate, often recurved . . . . . . . . . . . . . . . . . . . . . .8. *A. drummondii*

aa.  None of the leaves both cordate and petiolate.

  b.  Cauline leaves sessile, clasping the stem with auricled or cordate bases.

  c.  Phyllaries glandular; leaves entire or nearly so.

    d.  Ray flowers more than 40 and usually bright purple; plants glandular-pubescent in the inflorescence; leaves crowded . . . . . . . . . . . . . . . . . . . . . . . . . . . . . . . . . . . . . . . . . . . . . . . . . . . . . . . .11. *A. novae-angliae*

    dd.  Ray flowers fewer than 40, blue, pink, or lavender; plants glandular or not in the inflorescence.

      e.  Leaves strongly cordate-clasping, the auricled lobes making the leaves appear almost perfoliate . . . . . . . . . . . . . . . . . .12. *A. patens*

      ee.  Leaves slightly to markedly cordate-clasping, but leaves not appearing perfoliate . . . . . . . . . . . . . . . . . . . . . . . . . . . . .13. *A. oblongifolius*

  cc.  Phyllaries not glandular; leaves entire or serrate.

    d.  Leaves sharply serrate, acuminate, abruptly contracted into a broad petiolar wing, leaf base strongly auriculate-clasping; stems often zigzag; ray corollas pale blue or lavender. . . . . . . . . .14. *A. prenanthoides*

    dd.  Leaves serrate to entire, not abruptly narrowed as above, merely tapering at base; stems not zigzag.

      e.  Leaves linear to linear-lanceolate, tapering at tip, only slightly auriculate-clasping; stems slender; phyllaries with solid green midvein; plants of bogs . . . . . . . . . . . . . . . . . . . . . . .15. *A. junciformis* ee. Leaves wider, lanceolate to ovate-lanceolate, oblong or ovate.

  f. Stems glabrous, glaucous (obvious in the field); leaves glabrous . . . .
. . . . . . . . . . . . . . . . . . . . . . . . . . . . . . . . . . . . . . . . . . .16. A. *laevis*

  ff. Stems usually purple with spreading-hispid hairs, at least above; leaves scabrous to sub-scabrous above, hairy along midrib beneath. . . . . . . . . . . . . . . . . . . . . . . . . . . . .17. A. *puniceus*

bb. Leaves sessile or petiolate, not obviously clasping, or if slightly clasping then without cordate or auricled bases.

  c. Phyllaries rigid, margins of subulate green tips inrolled; ray corollas usually white; stems glabrous or pilose . . . . . . . . . . . . . .18. A. *pilosus*

  cc. Phyllaries flat, not inrolled, not subulate.

    d. Ray corollas seemingly absent (actually present in the form of a short tube without a ligule); leaves linear, glabrous except the ciliate margins; inflorescence open or somewhat spicate; rare. . . . . . . . . . . . . . .
. . . . . . . . . . . . . . . . . . . . . . . . . . . . . . . . . . . .19. A. *brachyactis*

    dd. Ray corollas obviously present, although ligule sometimes short.

      e. Ray corollas about the same length as the pappus bristles, rolled at the tip and more abundant than the disk flowers; leaves linear to lanceolate, entire; rare . . . . . . . . . . . . . . . . . . . . . . .20. A. *subulatus*

      ee. Ray corollas not as above, usually fewer than the disk corollas.

        f. Pappus in a double series, the outer shorter.

          g. Leaves linear, midrib strong, appearing 1-nerved; disk corollas red or yellowish; rays violet, rarely white; achenes with long white hairs. . . . . . . . . . . . . . . . . . . . . . . . . . . . .21. A. *linariifolius*

          gg. Leaves lanceolate or wider, obviously pinnately veined; rays white or creamy; inner pappus bristles clavate.

            h. Achenes glabrous; inflorescence open and spreading . . . . . . . . . .
. . . . . . . . . . . . . . . . . . . . . . . . . . . . . . . . . . . .22. A. *infirmus*

            hh. Achenes strigose or puberulent; inflorescence compact, a dense corymb . . . . . . . . . . . . . . . . . . . . . . . . . . . . .23. A. *umbellatus*

        ff. Pappus in a single series.

          g. Achenes with long silky hairs which sometimes appear as an outer pappus; ray corollas 3–8.

            h. Leaves linear to lanceolate, longitudinally 3-nerved, entire, glabrous; pappus white or pale . . . . . . . . . . . . . .24. A. *solidagineus*

            hh. Leaves lanceolate or lance-obovate, pinnately veined, nearly always dentate toward apex, somewhat scabrous; pappus reddish-brown when mature. . . . . . . . . . . . . . . . . . . . . . .25. A. *paternus*

          gg. Achenes sometimes pubescent but without long silky hairs; ray flowers more than 8.

x 1

x 1

x ½

Aster macrophyllus

x ½

Aster schreberi

h. Inflorescence corymbiform . . . . . . . . . . . . . . . .26. *A. surculosus*

hh. Inflorescence not corymbiform.

   i. Phyllaries mucronate, squarrose; heads many, small; inflorescence with many small bracts . . . . . . . . . . . . . . . . . . .27. *A. ericoides*

   ii. Phyllaries acute, not squarrose.

      j. Limbs of the disk corollas lobed 50–75% of the total length of the limb; inflorescences somewhat one-sided.

         k. Leaves lanceolate to lance-elliptic, scabrous above, glabrous beneath except on midvein; limb of the disk corollas goblet-shaped, the lobes recurving . . . . . . . . . . . .28. *A. lateriflorus*

         kk. Leaves linear to narrowly lanceolate, glabrous above and beneath; limb of the disk corollas funnel-shaped, the lobes ascending. . . . . . . . . . . . . . . . . . . . . . . . . . . .29. *A. vimineus*

      jj. Limbs of the disk corollas lobed less than 50% of the total length of the limb; inflorescences not one-sided.

         k. Leaves mostly entire, margins revolute, veins strongly reticulate beneath, areolae nearly isodiametric . . .30. *A. praealtus*

         kk. Leaves not as above, if reticulate beneath, areolae not isodiametric.

            l. Heads long-pedunculate; phyllaries with rhombic green tips; ray flowers blue or lavender; rare . . . . . . . . .31. *A. dumosus*

            ll. Heads short-pedunculate; green areas of phyllaries usually long, not rhombic; ray flowers white to pale blue.

               m. Leaves entire, linear to linear-ovate; rare plants of bogs . . . . . . . . . . . . . . . . . . . . . . . . . . . . . . . .15. *A. junciformis*

               mm. Leaves serrate; very common plants of moist places, but not bogs . . . . . . . . . . . . . . . . . . . . . . . . .32. *A. simplex*

## 1. **Aster macrophyllus** L.   LARGE-LEAVED ASTER

Stems glandular or viscid-glandular in the inflorescence, sometimes with spreading hairs, basal tufts abundant; leaves glabrous to scabrous, hirsute above to more villous beneath, crenate to more often serrate on margins, teeth mucronate, acuminate to obtuse, long-petiolate, cordate, especially the middle and lower; inflorescence corymbiform, bracts few; phyllaries glandular, ciliolate, often with short hairs, imbricate, sharp and acute to blunt; ray flowers pale to rather vivid lilac, or merely purple-tinged. Common in dry or moist woodland margins. July–Oct.

The many varieties described in the literature are puzzling, at least in Ohio, where they can hardly be separated and identified with certainty and consistency.

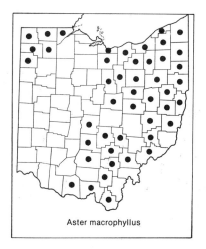

Aster macrophyllus

## 2. **Aster schreberi** Nees

Stems glabrous below, pilose in inflorescence and on peduncles, usually with tufts of basal or rosette leaves due to rhizomatous spreading; leaves rather large, usually with rectangular sinuses, margins usually sharply dentate-serrate, with mucronate teeth, middle and upper leaves becoming reduced and more narrow, often nearly sessile above; inflorescence corymbiform; phyllaries few, usually glabrous on back, lax and spreading, the inner longer, linear with inconspicuous greenish tips, blunt; ray flowers white. Rather common in moist woodlands and thin thickets. July–Oct.

Plants of this species are often mistakenly identified as A. *divaricatus* (see discussion under next species) or A. *macrophyllus*. The eglandular, narrow phyllaries separate them from plants of A. *macrophyllus*. There are many specimens, particularly in The Ohio State University Herbarium, that appear to be hybrids, presumably with A. *macrophyllus*.

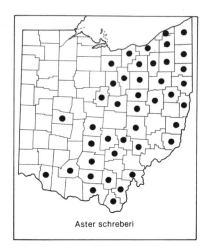

Aster schreberi

## 3. **Aster divaricatus** L.

Stems generally shorter than those of A. *schreberi*, slender, most often zigzag, glabrous or somewhat pubescent; leaves thin, glabrous above to slightly pubescent beneath, lower leaves cordate with rounded sinuses, sharply dentate margins, mucronate teeth, and slender petioles; inflorescence broad, corymbose; phyllaries imbricate, broad, blunt, ciliolate with inconspicuous green tips; ray flowers white. Rather common in open, dry woodland habitats. Aug.–Sept.

Most readily identified by the slender zigzag stems and leaves that are sharply and unevenly serrate.

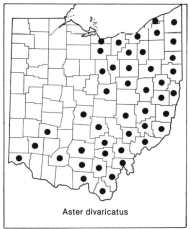

Aster divaricatus

## 4. **Aster undulatus** L.

Stems usually densely pubescent with short hairs or sometimes merely scabrous; leaves scabrous above, finely pubescent beneath, basal and lower cauline leaves long-petiolate, cordate, the middle and upper regularly reduced, petioles of mid-cauline leaves usually winged, upper leaves sessile, cordate or auriculate-clasping; inflorescence open, paniculate; phyllaries narrow, imbricate, minutely pubescent, tips rhombic; ray flowers blue or lilac. Dry, open woods and thickets. Aug.–Oct.

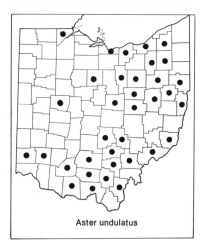

Aster undulatus

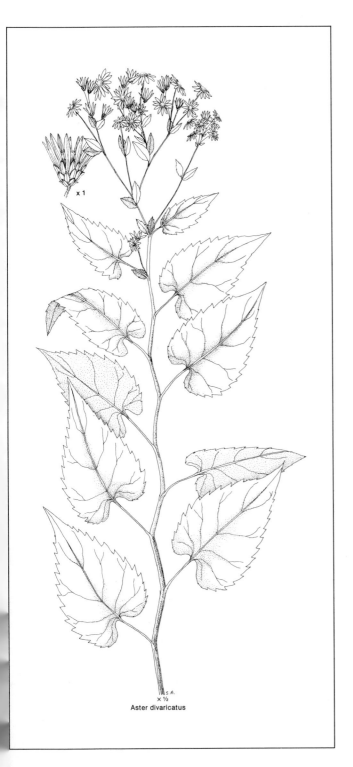

x 1

x ½
Aster divaricatus

x 2

x ½
Aster undulatus

## 5. **Aster shortii** Lindl.    SHORT'S ASTER

Stems essentially glabrous, most often striate; leaves nearly always entire, mostly glabrous above, with spreading pubescence beneath, nearly all leaves from base of plant to inflorescence petiolate and cordate, gradually reduced above; inflorescence paniculiform, bracteate; phyllaries narrow, finely pubescent, and with green area extending downward onto midvein; ray flowers usually blue, occasionally pink or white. Widely scattered in Ohio, often in dry woodland habitats. Aug.–Oct.

## 6. **Aster oolentangiensis** Riddell

　A. *azureus* Lindl.

　A. *capillaceus* Burgess

Stems scabrous to nearly glabrous; leaves entire to remotely toothed, scabrous above, merely pubescent beneath, lower leaves (often missing at anthesis) long-petiolate, cordate, ovate-lanceolate, acuminate, middle leaves with shorter petioles, lanceolate, tapering to petiole, upper leaves scant, becoming sessile, inflorescence open-paniculate; phyllaries linear with conspicuous diamond-shaped green area at tip, glabrous on back, ciliolate; ray flowers blue. Prairies and open, dry woodlands. July–Oct.

　　For justification of this recent name change for one of our common species, see Jones (1983).

## 7. **Aster sagittifolius** Willd.    ARROWLEAF ASTER

Stems mostly glabrous to minutely pubescent in lines; leaves cordate, shallowly serrate, glabrous (scabrous in ill-defined varieties); petiole winged, narrowly so in lower leaves; phyllary tips long, linear, and with narrow green stripe; ray flowers blue or lilac, rarely white. Widely distributed in Ohio. Common along stream banks and in woodlands, less common in open habitats. Aug.–Oct.

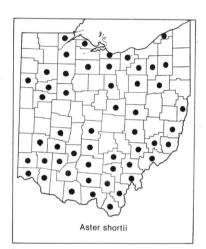

Aster shortii

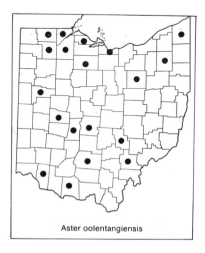

Aster oolentangiensis

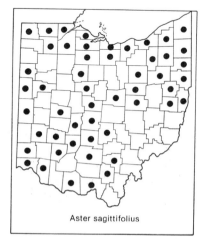

Aster sagittifolius

×1

×1

× ½
Aster shortii

× ½
Aster oolentangiensis

x 1

x 1

S.A.
× ½
Aster sagittifolius

× ½
Aster lowrieanus

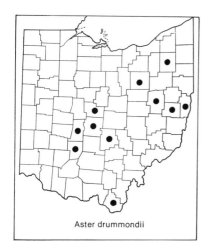

Aster drummondii

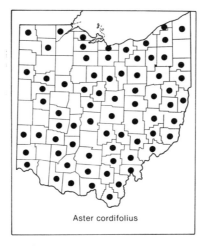

Aster cordifolius

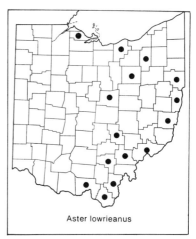

Aster lowrieanus

8. **Aster drummondii** Lindl.

Stems soft-pubescent, tending to be shorter than those of A. *sagittifolius*; leaves shallowly toothed, scabrous above, pubescent beneath, ovate to ovate-lanceolate, cordate; petiole wing very narrow; phyllaries linear (even longer than those of A. *sagittifolius*); the green tips elongate-rhombic, sometimes re-curved; ray flowers blue. Dry open woods and thickets in central, eastern, and southern counties; possibly locally abundant, but overall not common in Ohio. Sept.–Oct.

This species has been reduced by some authors to a variety of A. *sagittifolius*, with which it appears to intergrade. In Ohio, A. *drummondii* is distinct and in this treatment has been retained as a species. It differs from A. *sagittifolius* in being shorter in stature, in having soft, short pubescence on leaves and upper stems, and in having phyllaries that are longer and with a median green stripe that may be broader and more rhombic than in A. *sagittifolius*.

9. **Aster cordifolius** L.    Blue Wood Aster

Stems glabrous below, often pubescent in lines in inflorescence, and often striate; leaves broadly ovate, sharply toothed, acuminate, cordate, mostly scab-rous above, sparsely pubescent beneath with spreading hairs, petioles only slightly winged; inflorescence paniculate; phyllaries obtuse, green-tipped, the green coloring sometimes obscured by anthocyanin; ray flowers blue to bluish-purple. A common woodland species, often confused with A. *lowrieanus*. Aug.–Oct.

10. **Aster lowrieanus** T. C. Porter

Stems smooth and glabrous; leaves similar to those of A. *cordifolius* but smooth, glabrous, and somewhat narrower, petioles nearly always broadly winged; phyllaries linear-oblong with broad diamond-shaped green area at the tip; ray flowers blue-violet to roseate. This woodland species is uncommon in Ohio, found primarily in the nonglaciated Appalachian Plateau in the south-eastern and southern parts of the state. Sept.–Oct.

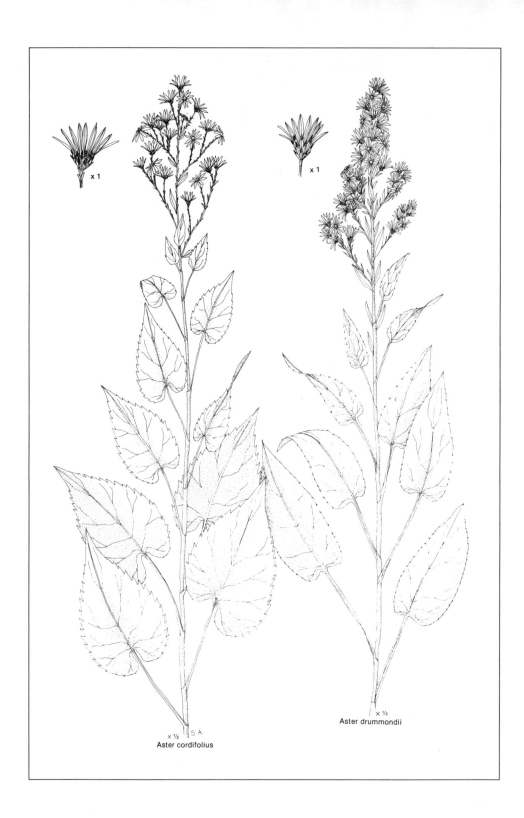

x 1

x 1

× ½
Aster drummondii

× ½ S. A.
Aster cordifolius

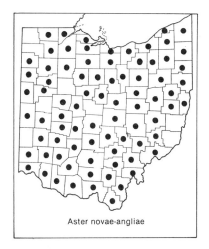

Aster novae-angliae

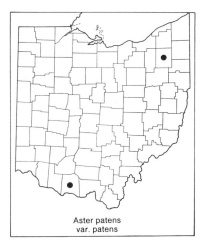

Aster patens
var. patens

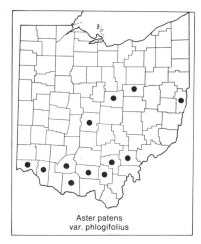

Aster patens
var. phlogifolius

Aster oblongifolius

### 11. **Aster novae-angliae** L.   New England Aster

Stems stout, pubescent, usually glandular upward, often forming large multi-stemmed clumps; leaves linear-lanceolate, often crowded, entire, sessile, clasping at base, scabrous; inflorescence showy; heads, peduncles, and phyllaries glandular; phyllaries slender, sometimes loose, spreading, tips acuminate, the outer somewhat foliaceous; ray flowers purple or rosy-purple, rarely white or merely blue. Open habitats, widely distributed in Ohio. Aug.–Oct.

### 12. **Aster patens** Ait.

Stems slender, short, pubescent; leaves thin, pubescent beneath, glabrous above, sessile, cordate and clasping, ovate to oblong, entire, lower leaves deciduous; inflorescence open, divaricately branched; phyllaries imbricated in several series, acute, short-hairy; ray flowers blue. Dry woods and dry open habitats. Aug.–Oct.

Two varieties can be recognized in Ohio, var. *patens* (illustrated) with thick and firm leaves rarely or only slightly constricted above the clasping base and the more common var. *phlogifolius* (Muhl.) Nees with thin leaves that are nearly always constricted above the clasping base.

### 13. **Aster oblongifolius** Nutt.   Aromatic Aster

Stems branched, glandular and hairy in branches; leaves entire, sessile, markedly to obscurely clasping, small, oblong to oblong-lanceolate, rarely 7 cm long, usually much smaller, scantly scabrous, lower leaves deciduous; inflorescences terminating branches; phyllaries glandular, few, green, loose and spreading, the tips acuminate; ray flowers blue to purple. Dry, open places in southern Ohio; rare. Aug.–Oct.

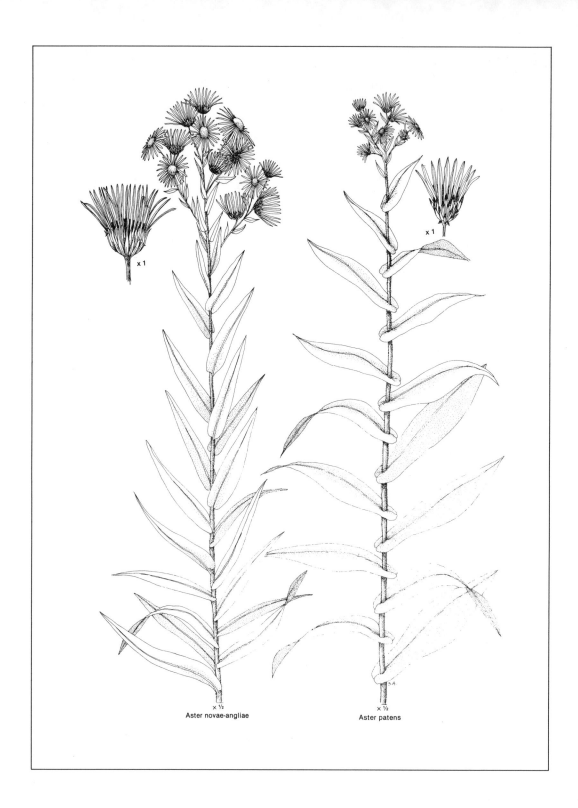

x 1

x 1

× ½

Aster novae-angliae

× ½

Aster patens

x 1

x 1

× ½
Aster oblongifolius

× ½
Aster prenanthoides

x 1

x 1

× ½
A
Aster junciformis

× ½
Aster laevis

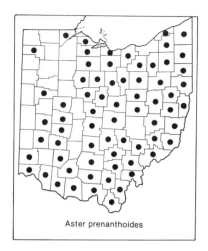

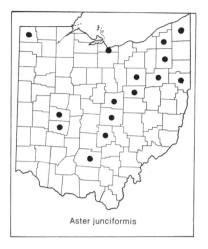

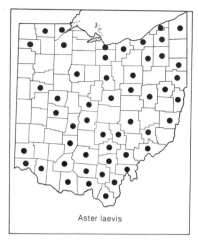

Aster prenanthoides          Aster junciformis          Aster laevis

14. **Aster prenanthoides** Muhl.  Crooked-stem Aster

Stems branched, often somewhat zigzag, pubescent in lines from leaf bases; leaves usually scabrous above, glabrous to pubescent beneath, ovate to ovate-lanceolate, serrate, acuminate, tapering rather abruptly into the winged petiole, auriculate-clasping at base, lower deciduous; heads numerous, inflorescence paniculate, open; phyllaries glabrous, loose, often spreading with age, the outer usually foliaceous; ray flowers blue or lavender. Widely distributed in Ohio where it occurs along stream banks, and in wet thickets, fields, and woodlands. Aug.–Oct.

15. **Aster junciformis** Rydb.

Stems slender, glabrous below, often pubescent in lines from petiole base, at least above; leaves linear, rarely 6–8 mm wide, sessile, often auriculate-clasping at base, entire, generally glabrous, midrib often pubescent below, lower leaves soon deciduous; heads few, even solitary in adverse habitats; phyllaries glabrous, slender, the outer green throughout, the inner with merely a solid green line, imbricate; rays white to pale blue, the lobes constituting about 20% of the corolla limb. Plants of bogs and wet places. Aug.–Oct.

16. **Aster laevis** L.  Smooth Aster

Stems glabrous; leaves glabrous (margins sometimes ciliolate to scabrous), most often glaucous (obvious in the field), entire or with a few remote teeth, but not sharply serrate, tapering to base and more or less auriculate-clasping; leaves reduced in inflorescence; inflorescence open, heads usually several, or only few in dry habitats; phyllaries imbricate in several series, their tips diamond-shaped; ray flowers blue or purplish; pappus often reddish in recently collected specimens and in the field. Open, dry habitats; common. Aug.–Oct.

### 17. **Aster puniceus** L.

Stems stout, often much-branched and with spreading hairs above to nearly glabrous below; leaves sessile, auriculate-clasping, although often only slightly so, subserrate to entire, usually scabrous above, pubescent along midrib to nearly glabrous beneath; inflorescence frequently leafy; heads few to many; phyllaries slender-acuminate, loose; ray flowers blue, occasionally white. Common in swamps, bogs, and wet stream valleys. Aug.–Nov.

Specimens assignable to var. *firmus* (Nees) T. & G. can occasionally be recognized in Ohio, and var. *lucidulus* Gray appears to be represented by some specimens in The Ohio State University Herbarium, but neither variety is sufficiently well-defined to warrant recognition. Both varieties are here included within the species.

### 18. **Aster pilosus** Willd.

Stems often much-branched, especially in the inflorescence, closely pilose to glabrous; leaves hirsute to glabrous, the lower soon deciduous, the upper linear to lance-elliptic, becoming reduced upward to mere bracts in the inflorescence; heads usually quite numerous, often secund; phyllaries imbricate, subulate, their tips marginally inrolled, green; ray flowers white, rarely pink or purplish. Very common in Ohio, where it occurs in old fields and waste places. Aug.–Oct.

For a study of the cytogeography, see Semple (1978).

In addition to var. *pilosus*, two others can be recognized among Ohio plants, var. *pringlei* (Gray) Blake and var. *demotus* Blake. The var. *pringlei* is common in the northern lake counties, although I have also seen a few specimens from Adams, Scioto, and Vinton Counties in southern Ohio. The var. *demotus*, which has often been confused with *A. ericoides*, has a scattered distribution in Ohio. The var. *pilosus* is widespread in the state. I consider var. *platyphyllus* (T. & G.) Blake to be merely an environmental form of var. *pilosus*.

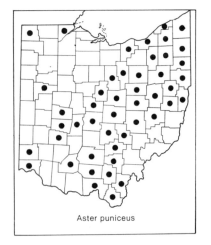

Aster puniceus

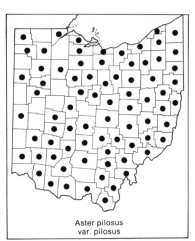

Aster pilosus
var. pilosus

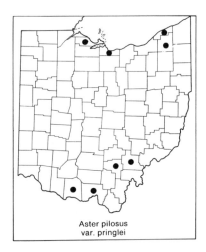

Aster pilosus
var. pringlei

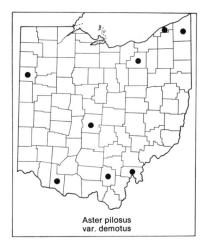

Aster pilosus
var. demotus

x 1

x 1

× ½

Aster puniceus

× ½

Aster pilosus
var. pilosus

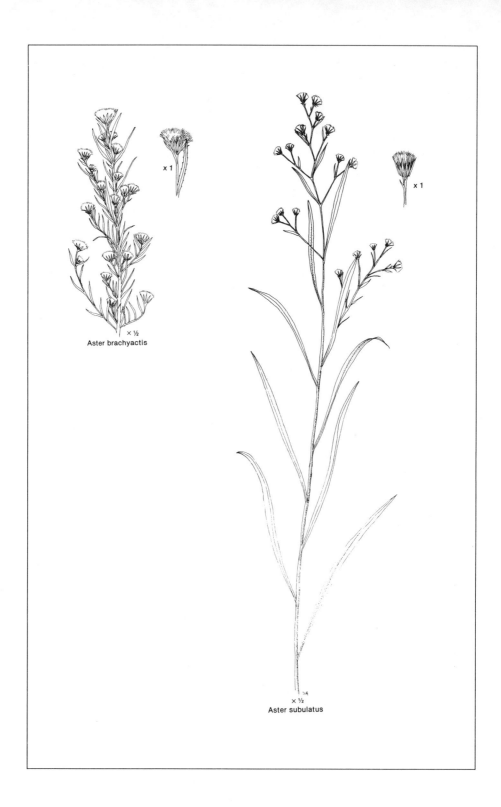

× ½
Aster brachyactis

x 1

x 1

× ½
Aster subulatus

a. Stems and leaves densely hirsute......................var. *pilosus*

aa. Stems and leaves glabrous.

b. Stems short, rarely over 30 cm tall; inflorescence subcorymbose-paniculate with ascending branches....................var. *pringlei*

bb. Stems taller, exceeding 30 cm; inflorescence loosely ascending, racemose-paniculate..............................var. *demotus*

19. ASTER BRACHYACTIS Blake

Stems often much-reduced, glabrous, short (at least in Ohio); leaves linear, entire, ciliolate, and crowded, the lower soon deciduous; inflorescence somewhat spiciform to short-paniculate; phyllaries loose, linear, acute, the outer somewhat herbaceous; marginal flowers rayless, reduced to a tube much shorter than the style, or the ligule a mere rudiment. Moist, usually saline, waste ground. Aug.–Oct.

Introduced from western North America, this species is represented in Ohio by a single collection from Ottawa County deposited in The Ohio State University Herbarium. While there is no doubt concerning the identification of this specimen, the continued existence of this species in Ohio is doubtful.

20. ASTER SUBULATUS Michx.

Stems glabrous; leaves linear to linear-elliptic or narrowly oblong, entire, glabrous; heads arranged in open panicles; phyllaries imbricated, acuminate, the inner with scarious, hyaline margins; ray corollas short, outwardly circinately rolled; pappus copious, white, conspicuous on older heads. Saline habitats. July–Sept.

A very rare species represented chiefly by a few collections of E. L. Moseley, made in 1932 around old oil well sites in Wood and Lucas Counties. I have recently searched these areas and found no specimens of this species. As with *A. brachyactis*, its continued occurrence in Ohio is doubtful. Roberts and Cooperrider (1982) consider it to be nonindigenous to Ohio.

Aster brachyactis

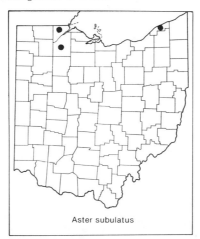

Aster subulatus

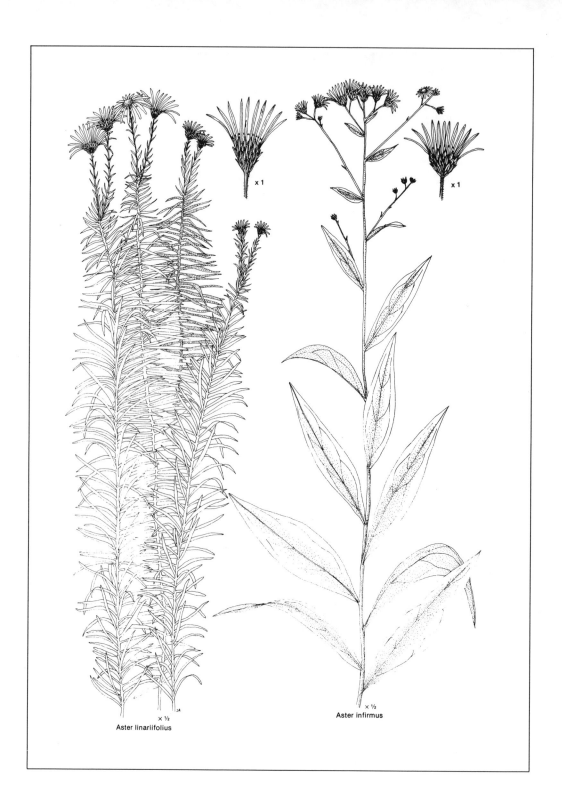

×1

×1

× ½
Aster linariifolius

× ½
Aster infirmus

### 21. **Aster linariifolius** L.

Stems somewhat wiry, usually several from a common caudex, pubescence of short hairs; leaves numerous, often crowded, linear, entire, ciliolate along margins, minutely scabrous, with lateral veins scarcely apparent, lower leaves soon deciduous; inflorescence usually somewhat corymbose, heads sometimes solitary; phyllaries imbricate, acute to rounded, with green area along keel; ray corollas blue to bluish-violet; pappus double. A very distinct but uncommon species in Ohio, occurring only in a few south-central counties in dry woods and dry, open places. Aug.–Oct.

### 22. **Aster infirmus** Michx.

Stems glabrous, plants without rhizomes; leaves ovate-elliptic, entire, mostly glabrous above, most often short-hirsute on veins beneath, sessile or nearly so, lower leaves soon deciduous; inflorescence somewhat corymbiform; phyllaries imbricate, broad, green or green-tipped; ray flowers white, showy; pappus double, the inner clavate, longer than the outer. Rather rare in Ohio, where it occurs in dry woodlands. July–Sept.

### 23. **Aster umbellatus** Mill.

Plants with creeping rhizomes; stems several from caudex, glabrous below the inflorescence; leaves narrow-elliptic to elliptic-ovate, short-petiolate to subsessile, entire, scabrous above, mostly glabrous beneath, lower leaves early deciduous; inflorescence corymbiform, often flat-topped; phyllaries mostly glabrous, imbricate, green-tipped; ray flowers white; pappus double. Wet meadows and stream bottoms. July–Sept.

### 24. **Aster solidagineus** Michx.    Narrow-leaved Aster

    *Sericocarpus linifolius* (L.) B.S.P.

Stems glabrous, several from thick caudex; leaves entire, linear or linear-

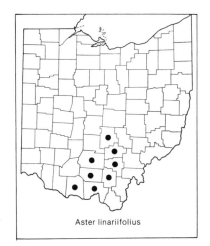

Aster linariifolius

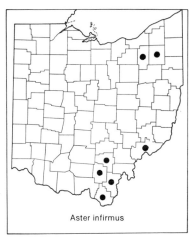

Aster infirmus

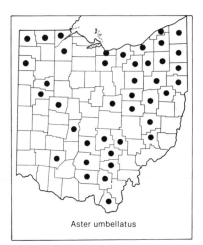

Aster umbellatus

Aster solidagineus

x 1

x 1

× ½
Aster umbellatus

× ½
Aster solidagineus

oblong, sessile, at least above; heads arranged in a flat-topped corymb; phyllaries broad, firm, the outer with green tips, the inner erose; rays 3–6, white; achenes silky. A rare aster in Ohio, known from dry, open places and open woods in Adams, Jackson, Pike, and Scioto Counties. June–Aug.

25. **Aster paternus** Cronq.

Sericocarpus asteroides (L.) B.S.P.

Stems glabrous below, glabrous or somewhat puberulent in inflorescence; leaves ciliolate along margins, often pubescent on surfaces, lower oblanceolate to elliptic, petiolate, becoming sessile upward on stem, teeth nearly always restricted to the terminal one-third of the leaf; heads arranged in a flat-topped corymb, glomerulate; phyllaries imbricate, glabrous, outer green-tipped; ray flowers 4–8, white; achenes densely sericeous, the outer hairs sometimes appearing as an outer pappus. Dry woodlands; common in unglaciated regions of Ohio. June–Sept.

26. **Aster surculosus** Michx.   Creeping Aster

Stems spreading-hairy, scantly glandular and viscid, especially below the inflorescence. A distinct species, similar in morphology to a more coastal species, A. spectabilis Ait. Dry woods and openings. Aug.–Oct.

Although a rare species represented in Ohio by a single collection from Scioto County, it is included here due to the close affinity of the flora of the collection site to that of the natural range of the species, i.e., the Appalachian region of Kentucky and North Carolina.

27. **Aster ericoides** L.   White Heath Aster

Stems pubescent, usually well-branched; leaves abundant, linear, sessile, the middle and lower leaves soon deciduous, the upper often reduced to bracts; heads many, often secund on divergent branches; phyllaries imbricate, mostly

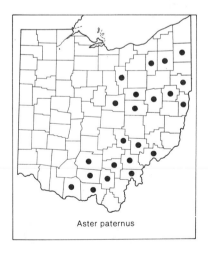

Aster paternus

Aster surculosus

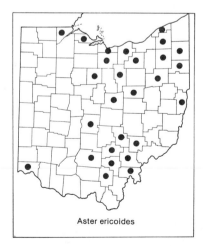

Aster ericoides

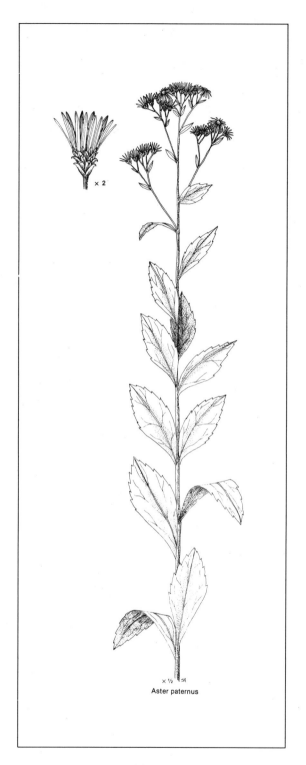

× 2

× ½

Aster paternus

× ½

Aster surculosus

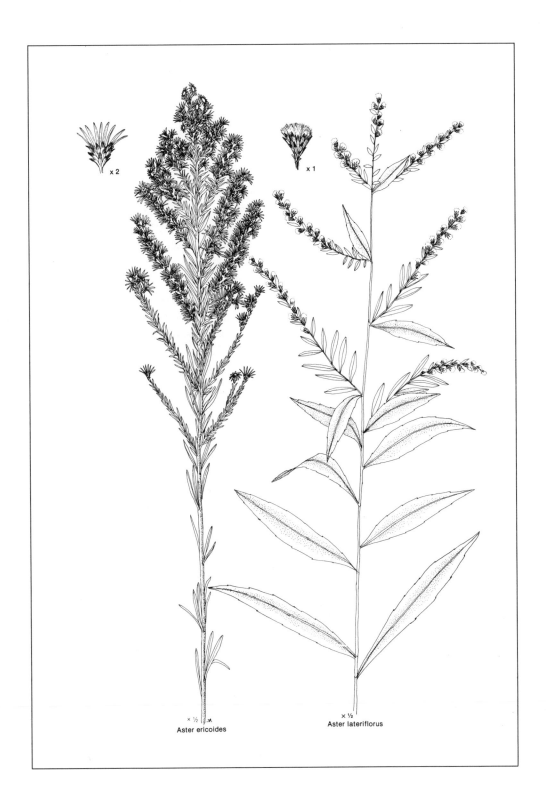

x 2

x 1

× ½

Aster ericoides

× ½

Aster lateriflorus

acute, spinulose-mucronate, somewhat squarrose, often ciliolate, pubescent on back; ray flowers white or pinkish; achenes sericeous. Dry, open places and thickets. Aug.–Sept.

Variety *prostratus* (Kuntze) Blake, with spreading-hairy stems, has been reported from the lake counties, but its existence in Ohio today is doubtful.

**28. Aster lateriflorus** (L.) Britt.   STARVED ASTER

Stems tall, glabrous to rather densely villous; leaves scabrous to nearly glabrous above, pubescent on veins beneath, lower leaves deciduous at anthesis, short-petiolate, mid and upper essentially sessile, linear-lanceolate or lance-elliptic, tapering from about the middle to the apex and base, entire or serrate; phyllaries glabrous to minutely pubescent on back, imbricate with broad, green tips most often suffused with purple near apex; ray flowers white or tinged with pink. Common in open woodlands and fields. Aug.–Oct.

**29. Aster vimineus** Lam.

Stems glabrous, or occasionally pubescent in lines; leaves glabrous above and beneath, linear to lanceolate, tapering to sessile base, entire to remotely toothed, much reduced in inflorescence; inflorescence open, paniculate, divaricate, the branches often recurved, often secund; phyllaries glabrous, imbricate, the tips elongated, green; ray flowers white. Moist open places, mostly river bottoms; scattered throughout eastern and southern Ohio. Aug.–Oct.

**30. Aster praealtus** Poir.

Stems usually pubescent in lines, particularly upward, often reddish-brown; leaves conspicuously reticulate beneath, the areolae nearly isodiametric, leaf margins entire or with a few teeth, blades glabrous or nearly glabrous above, sessile; heads paniculately arranged; phyllaries essentially glabrous, imbricate, narrow; ray flowers blue to bluish-purple. Low moist habitats; scattered distribution in Ohio. Sept.–Oct.

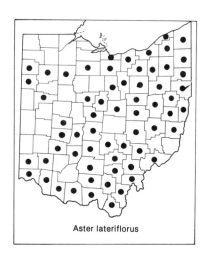

Aster lateriflorus

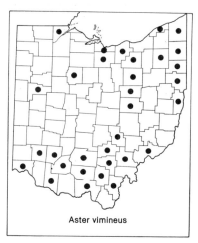

Aster vimineus

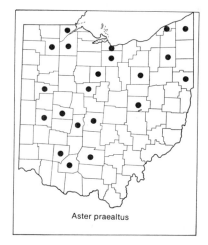

Aster praealtus

× 3

× 1

× ½
Aster vimineus

× ½
Aster praealtus

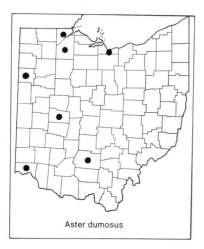

Aster dumosus

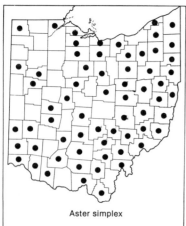

Aster simplex

### 31. **Aster dumosus** L.

Stems glabrous or pubescent in lines; leaves linear to lanceolate, sessile, entire to remotely toothed, slightly scabrous above, glabrous beneath, reduced in inflorescence branches; inflorescence open, usually much-branched, heads long-pedunculate, subtended by numerous small leaves or bracts; phyllaries imbricate, glabrous, broad with green tips; ray flowers blue to bluish-lavender. Moist habitats, usually swampy places. Not well-represented in the Ohio flora, but plentiful in Champaign County at Cedar Swamp and in Wood County, where it occurs in low, wet woodlands. Aug.–Oct.

### 32. **Aster simplex** Willd.

Stems tall, glabrous below, most often pubescent in lines above; leaves linear to lanceolate, serrate to subentire, usually glabrous on both surfaces, tapering to petiole-like base, veinlets beneath obscurely reticulate, if visible, the areolae irregular; inflorescence usually leafy; phyllaries acute, glabrous, margins ciliolate, imbricate, the green tips elongated; ray flowers usually white. Moist areas of ravines, river bottoms, and wet woodlands. Aug.–Oct.

Three varieties can be recognized in Ohio: var. *simplex*, var. *ramosissimus* (T.&G.) Cronq. (illustrated), and var. *interior* (Weig.) Cronq.

a. Heads small, 3 mm or less in height . . . . . . . . . . . . . . . . . . . .var. *interior*
aa. Heads larger, exceeding 4 mm.
  b. Largest leaves 2–4 cm wide . . . . . . . . . . . . . . . . . . . . . . . . .var. *simplex*
  bb. Largest leaves less than 1.5 cm wide. . . . . . . . . . . . . . .var. *ramosissimus*

### 16. Conyza L.   HORSEWEED

Heads many, small, usually 5 mm or less in width. Ray flowers with short, narrow and inconspicuous white to purplish corollas about the same length as the disk flowers. Disk flowers rarely exceeding 20. Receptacles flat, naked. Achenes with 1 or sometimes 2 faint nerves, or nerveless. Pappus of capillary bristles, sometimes with an additional row of short bristles. Weedy, annual herbs with a rather well-developed taproot.

Plants of *Conyza* are often difficult to separate from those of the genus *Erigeron*, but the much smaller head size is a consistent differentiating character. Composed of 40–50 species, *Conyza* is mostly tropical and subtropical. The following two species have become naturalized in Ohio:

a. Stems unbranched to the inflorescence; ray flowers white . . . . . . . . . . . .
  . . . . . . . . . . . . . . . . . . . . . . . . . . . . . . . . . . . . . . . . . . . . . . . . .1. *C. canadensis*
aa. Stems branched from the base; ray flowers purplish . . . .2. *C. ramosissima*

x 1

x 1

× ½
**Aster dumosus**

× ½
**Aster simplex**

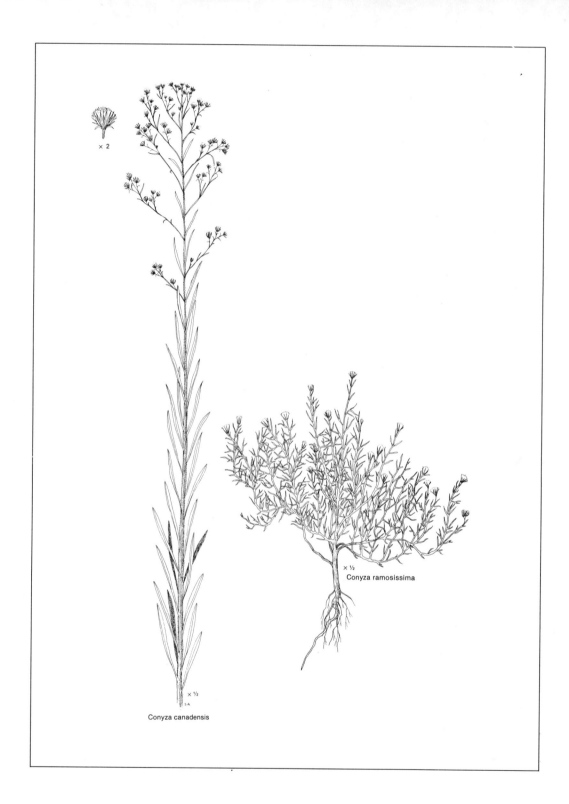

× 2

× ½
Conyza ramosissima

× ½
S A

Conyza canadensis

1. Conyza canadensis (L.) Cronq.  Common Horseweed

   *Erigeron canadensis* L.

Annual weed, 0.5–1 m tall; stems spreading-hirsute to nearly glabrous, un-branched to the inflorescence; leaves linear to oblanceolate, dentate to entire, reduced upward, lower leaves often deciduous by anthesis; heads many, small, less than 5 mm wide, in open paniculate arrangement; phyllaries imbricate, the midvein brownish; ray flowers barely equaling the disk flowers, white, rarely pinkish. Common weed in waste places and cultivated fields, and along roadsides throughout North America. July–Nov.

   Two varieties can be recognized in eastern United States; our plants, with stems spreading-hirsute, mostly belong to var. *canadensis*. A single specimen in The Ohio State University Herbarium, collected from Gallia County, is repre-sentative of var. *pusillus* (Nutt.) Ahles, in which the stems are glabrous or nearly so.

2. Conyza ramosissima Cronq.

   *Erigeron divaricatus* Michx.

Low annual with stems barely 10 cm tall, diffusely branched; leaves narrowly linear, rarely to 4 cm long, the upper reduced to bracts; heads many, small, much like those of *C. canadensis*; phyllaries imbricate, acute, the outer pubes-cent, the inner glabrous; ray flowers small, purplish. Weedy, but not common where it occurs in sandy soil of waste places and along streams. June–Sept.

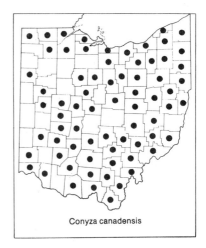

Conyza canadensis

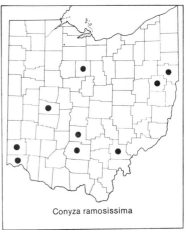

Conyza ramosissima

## 17. Erigeron L.  DAISY FLEABANE

Heads radiate, at least 1 cm wide. Phyllaries herbaceous and equal to scarcely herbaceous and imbricate. Receptacle flat, naked. Disk flowers yellow, numer-ous, perfect, seed-forming. Achenes faintly nerved, usually minutely pubes-cent. Pappus of capillary bristles, often minute and fragile, with an additional row of short bristles or scales sometimes present. Fibrous-rooted annuals, per-ennials, or biennials.

   A very large genus of at least 200 species distributed world-wide. The genus is often difficult to distinguish from *Conyza*, which is sometimes included within *Erigeron*. In our range, the head width of at least 1 cm distinguishes *Erigeron* from the small-headed *Conyza*. The present treatment follows that of Cronquist (1947).

a.  Cauline leaves narrowed to the base; pappus of ray flowers absent or of short scales or bristles; rays mostly white.

b.  Hairs of the stem long and spreading; basal leaves mostly elliptic to ovate, to 6 cm wide; cauline leaves broadly lanceolate, usually sharp-toothed . . . . . . . . . . . . . . . . . . . . . . . . . . . . . . . . . . . . . . . .1. *E. annuus*

bb. Hairs of the stem short and mostly appressed; basal leaves oblanceolate, to 2.5 cm wide; cauline leaves linear to narrowly lanceolate, entire to toothed . . . . . . . . . . . . . . . . . . . . . . . . . . . . . . . . . . . . . . . .2. *E. strigosus*

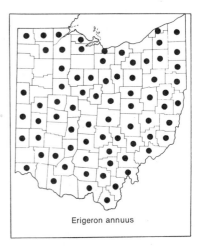

Erigeron annuus

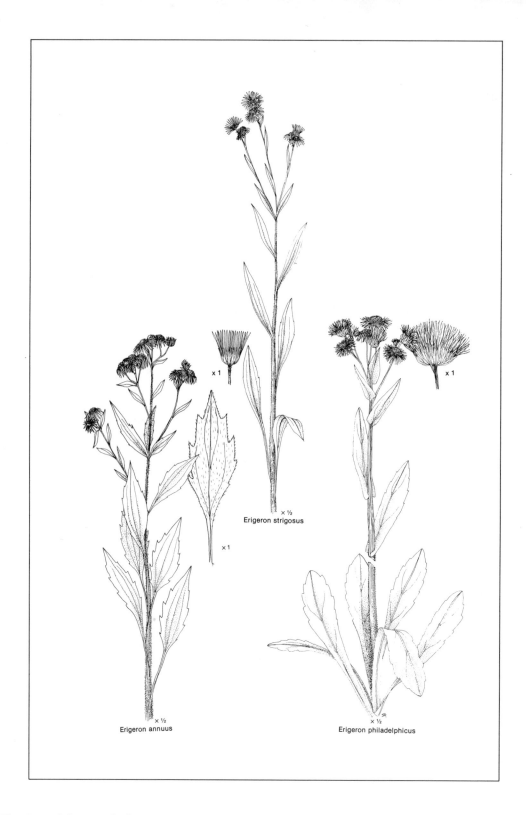

x 1

x 1

Erigeron strigosus

x 1

x 1

Erigeron annuus

Erigeron philadelphicus

aa. Cauline leaves clasping or sessile; pappus of ray flowers capillary.

    b.  Ray flowers 100 or more, pink to white; heads less than 2.5 cm wide . . . . . . . . . . . . . . . . . . . . . . . . . . . . . . . . . . . . . . . . . . . . . . . . . . .3. *E. philadelphicus*

    bb. Ray flowers fewer than 100, blue, violet, or white; heads 2.5 cm or more wide . . . . . . . . . . . . . . . . . . . . . . . . . . . . . . . . . . . . . . . .4. *E. pulchellus*

### 1. **Erigeron annuus** (L.) Pers. DAISY FLEABANE. WHITE-TOP

Annual or more often biennial from first year rosettes; stems to 0.5 m tall, hirsute with spreading hairs except in inflorescences; basal leaves elliptic to suborbicular to ovate, coarsely toothed, long-petiolate, mid cauline leaves lanceolate to barely ovate-lanceolate, rather sharply toothed; heads usually numerous; phyllaries often glandular; ray flowers white to pinkish; achenes faintly 2-nerved; pappus of disk flowers double, of fragile bristles and short scales; pappus of ray flowers of scales only. Weed of waste places, common to all counties. June–Oct.

### 2. **Erigeron strigosus** Muhl. ex Willd. WHITE-TOP

Annual or frequently a biennial forming rosettes; stems to 0.5 m tall, short-appressed hairy, sometimes spreading-hairy toward base; leaves mostly entire, occasionally toothed, linear to narrowly lanceolate; heads several; phyllaries obscurely glandular and short-hairy; ray flowers many, 50–100, white or pinkish; pappus of ray flowers of scales only. Common weed of waste places, fields, and roadsides. May–Sept.

### 3. **Erigeron philadelphicus** L. PHILADELPHIA FLEABANE

Apparently biennial; stems to 1 m tall, subglabrous to pubescent; cauline leaves clasping at base, ovate to ovate-oblong, reduced upward, basal leaves obovate to oblanceolate, crenate to completely entire; heads few to many; phyllaries hirsute to subglabrous; ray flowers many, 100 or more, white to pink or rose; achenes faintly 2-nerved; pappus of simple bristles. Common in a variety of habitats, often weedy. April–Aug.

### 4. **Erigeron pulchellus** Michx. ROBIN'S PLANTAIN

Perennial or biennial from rosettes formed in late fall; stem villous, especially below, less branched and with fewer heads than *E. philadelphicus*; leaves villous, quickly reduced upward, ovate to ovate-lanceolate, spatulate below, crenate or obscurely dentate; heads on long peduncles, solitary to only a few; phyllaries imbricate; heads usually 2.5 cm or more wide; ray flowers 50–100; achenes 1–2-nerved or nerveless; pappus of simple capillary bristles. Common in open woods, fields, and roadsides. April–July.

Two varieties may be recognized, var. *pulchellus* (illustrated), the most common and widespread, and var. *brauniae* Fern. which occurs in a few southern counties. The latter can be detected by its essentially glabrous stems and leaves.

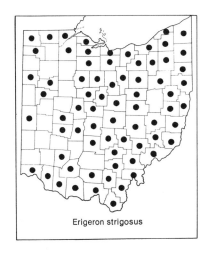

Erigeron strigosus

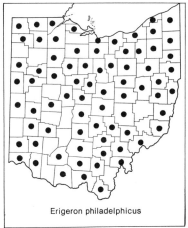

Erigeron philadelphicus

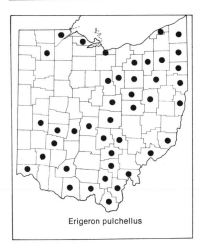

Erigeron pulchellus

## TRIBE IV. INULEAE CASS.

  a.  Heads radiate, the rays yellow; cauline leaves clasping . . . . . . . .23. *Inula*

aa.  Heads discoid, small; cauline leaves not clasping.

    b.  Young stems and leaves flocculent-tomentose, lower leaf surfaces with short, thick pubescence, giving the appearance of felt.

      c.  Receptacles chaffy, a scale subtending each achene; plants annual . . . .
. . . . . . . . . . . . . . . . . . . . . . . . . . . . . . . . . . . . . . . . . . . . . . . . . . . . . .19. *Filago*

    cc.  Receptacles naked.

Erigeron pulchellus
× ½

Pluchea camphorata
× ½

d. Plants dioecious, with all heads carpellate or all staminate, never both; plants usually forming rosettes at anthesis . . . . .20. *Antennaria*

dd. Plants rarely or only imperfectly dioecious, with heads of perfect or imperfect flowers or a mixture of the two; plants rarely forming rosettes at anthesis.

e. Heads with both perfect and imperfect flowers, the outer flowers carpellate and with filiform corollas, a few inner flowers perfect and with coarser 5-toothed corollas. . . . . . . . . . . . . . . .22. *Gnaphalium*

ee. Heads with imperfect flowers only; carpellate flowers with filiform corollas, staminate flowers with corollas coarse and lobed; sometimes a few staminate flowers found near center of carpellate heads. . . . . .
. . . . . . . . . . . . . . . . . . . . . . . . . . . . . . . . . . . . . . .21. *Anaphalis*

bb. Young stems and leaves without flocculent tomentum, lower leaf surfaces not felted; plants somewhat glandular, emitting a strong unpleasant camphoric odor . . . . . . . . . . . . . . . . . . . . . . . . . . . . . . . . . . . . . .18. *Pluchea*

## 18. Pluchea Cass. MARSH-FLEABANE

Heads many-flowered, discoid, pink to white, outer flowers carpellate and seed-forming, corollas filiform, inner flowers tubular, apparently perfect but rarely setting seed. Phyllaries imbricate, not scarious. Receptacle flat, naked. Anthers with tails. Achenes 4–5-angled. Pappus in a single row of numerous rough bristles. Herbs with alternate, toothed, ovate to elliptic leaves.

1. **Pluchea camphorata** (L.) DC. MARSH-FLEABANE

Wet places in open woods. Aug.–Sept.

Known only from Brown County specimens collected in 1923, the species is probably extirpated from Ohio (Roberts and Cooperrider, 1982). In the basic taxonomic treatment for the genus, Godfrey (1952) cites a specimen, which I have not seen, from Hamilton County.

Pluchea camphorata

## 19. Filago L. COTTON-ROSE. CUDWEED

Heads discoid, numerous, whitish, aggregated in small, dense, sessile clusters which are usually subtended by leafy bracts. Phyllaries small, imbricate, scarious, the outer terminating in a short bristle. Outer flowers carpellate, seed-forming, subtended by boat-shaped bracts. Inner flowers without subtending bracts, perfect but rarely forming seeds. Pappus capillary, present in inner flowers only. Stems woody, usually several from the base, usually branched. Leaves linear, alternate, entire, white-woolly.

1. FILAGO GERMANICA (L.) Huds. COTTON-ROSE

Naturalized from Europe. In dry fields, not widespread in Ohio. May–Sept.

Filago germanica

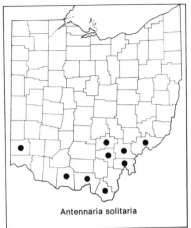

Antennaria solitaria

## 20. Antennaria Gaertn. PUSSY-TOES. EVERLASTING

Dioecious species. Heads discoid, solitary or in close clusters. Phyllaries imbricate in several series, scarious, often with petaloid tips. Receptacle flat or slightly convex, naked. Staminate plants usually smaller than carpellate plants. Staminate flowers usually with reduced pappus, commonly of barbellate or clavate bristles. Carpellate flowers with naked capillary bristles. Achenes terete or nearly so. Perennial, woolly, usually stoloniferous herbs with alternate and basal leaves.

Often a very confusing genus due in part to apomixis and polyploidy which are widespread. Staminate plants are usually absent from populations. This treatment follows, in part, that of R. J. Bayer and G. L. Stebbins (1982). A cytotaxonomic work such as this was long needed for this complex taxon. In Ohio, three sexual diploid species (*A. solitaria*, *A. virginica* and *A. neglecta*) and two polyploid complexes (*A. parlinii* and *A. neodioica*) can be recognized. The

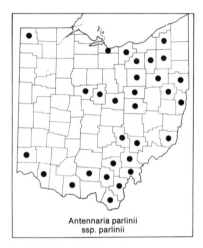

Antennaria parlinii
ssp. parlinii

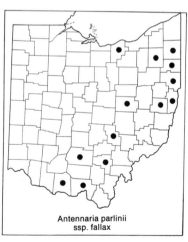

Antennaria parlinii
ssp. fallax

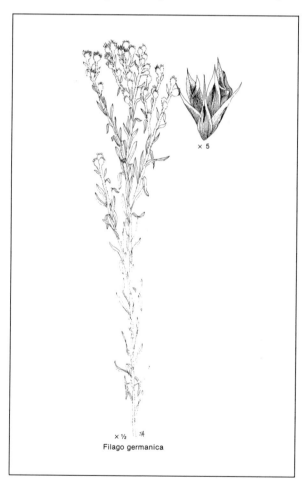

× 5

× ½

Filago germanica

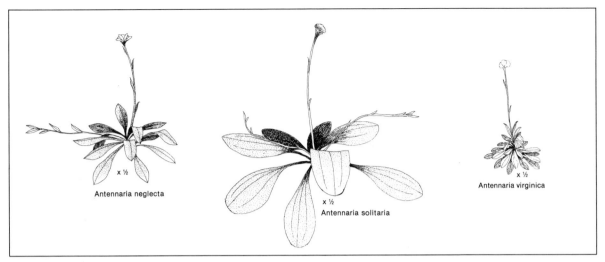

Antennaria neglecta   x ½

Antennaria solitaria   x ½

Antennaria virginica   x ½

cytology of the eastern species of *Antennaria* is presented by Bayer and Stebbins (1981).

a. Heads solitary .....................................1. *A. solitaria*
aa. Heads 2 or more per flowering stalk.
 b. Rosette leaves with 3–8 nerves .......................2. *A. parlinii*
 bb. Rosette leaves prominently 1-nerved.
  c. Upper cauline leaves with flat or curled, awn-like tip; involucral bracts usually brown at base ...........................3. *A. neglecta*
  cc. Upper cauline leaves without flat awn-like tips.
   d. Carpellate plants common, staminate plants absent or rare; rosette leaves longer than 2.5 cm, cauline leaves wider than 2 mm........
   ...........................................4. *A. neodioica*
   dd. Carpellate and staminate plants about equally common in the populations; leaves much smaller than the above, rosette leaves rarely 1.5 cm long and 2 mm wide; rare in Ohio................5. *A. virginica*

### 1. **Antennaria solitaria** Rydb.   SINGLE-HEADED EVERLASTING

Staminate and carpellate plants similar; stems extensively creeping with prostrate branches and long leafy stolons terminating in rosettes; rosette leaves obovate-spatulate to oblong-spatulate, 3–5-nerved, sessile or commonly with short, broad petioles, tomentose; flowering stems with narrow bract-like leaves; heads solitary, terminal, rather large and showy, brownish when dry; phyllaries linear, the tips thin. Woodlands and clearings. April–June.

Antennaria neglecta

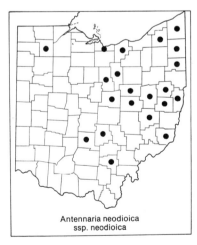

Antennaria neodioica
ssp. neodioica

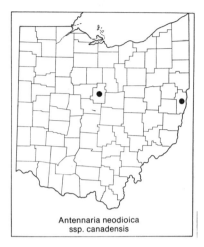

Antennaria neodioica
ssp. canadensis

Antennaria neodioica
ssp. petaloidea

Antennaria virginica

2. **Antennaria parlinii** Fern.   Parlin's Everlasting

Stems green or purplish, glandular or glandless; rosette leaves glabrous or tomentose above, bright green, obovate to obovate-spatulate, petiolate, and with prominent mucronate tips, 3–8-nerved; heads large, phyllaries lanceolate. Dry fields and waste places. April–July.

According to Bayer and Stebbins (1982) this species consists of two subspecies, ssp. *parlinii* and ssp. *fallax* (Greene) Bayer & Stebbins.

a.   Basal leaves glabrous above; young stems glandular at summit. . . . . . . . . . . . . . . . . . . . . . . . . . . . . . . . . . . . . . . . . . . . . . . . . . . . . . . . . . . . . . . . . . . .ssp. *parlinii*
A. *arnoglossa* Greene, A. *plantaginifolia* (L.) Richards var. *arnoglossa* (Greene) Cronq.

aa.  Basal leaves tomentose above; young stems usually glandless. . . . . . . . . . . . . . . . . . . . . . . . . . . . . . . . . . . . . . . . . . . . . . . . . . . . . . . . . . . . . . . . . . . . . . . . . . . . . . . . .ssp. *fallax*
A. *fallax* Greene, A. *arnoglossa* Greene var. *ambigens* Greene, A. *parlinii* Fern. var. *ambigens* (Greene) Fern., A. *ambigens* (Greene) Fern., A. *plantaginifolia* (L.) Richards var. *ambigens* (Greene) Cronq., A. *farwellii* Greene

3. **Antennaria neglecta** Greene   Field Everlasting

Stems to 40 cm tall, spreading by flexuous stolons which develop rosettes at their tips; rosette leaves oblanceolate to spatulate, 1-nerved, narrowed at base to short petiole, canescent above; flowering stems slender, elongated in fruit, staminate plants soon withering, cauline leaves 3–10 with flat, curled, awn-like tips. Common in sterile fields and pastures. Midspring–early summer.

4. **Antennaria neodioica** Greene   Smaller Everlasting

Basal stolons many, soon producing rosettes; rosette leaves obovate, mucronate-tipped, petiolate, gray-tomentose or bright green and essentially glabrous above; flowering stems slender; cauline leaves often sparse and irregularly spaced; staminate plants rare. Dry open places. Midspring–late summer.

Interpreted by Bayer and Stebbins (1982) as consisting of three subspecies, namely, ssp. *neodioica*, ssp. *canadensis* (Greene) Bayer & Stebbins, and ssp. *petaloidea* (Fern.) Bayer & Stebbins.

a.   Young leaves glabrous and bright green above . . . . . . . . . . .ssp. *canadensis*
aa.  Young leaves tomentose and gray-green above.
  b.   Basal leaves with distinct petioles, all leaves about the same size, mid cauline leaves 2 mm or more wide . . . . . . . . . . . . . . . . . . .ssp. *neodioica*
  bb.  Basal leaves gradually tapering to base, those along stolons smaller than rosette leaves, mid cauline leaves rarely 1 mm wide. . . . . . .ssp. *petaloidea*

5. **Antennaria virginica** Stebbins
   incl. A. *virginica* Stebbins var. *argillicola* Stebbins,
   A. *neodioica* Greene var. *argillicola* (Stebbins) Fern.,
   A. *neglecta* Greene var. *argillicola* (Stebbins) Cronq.

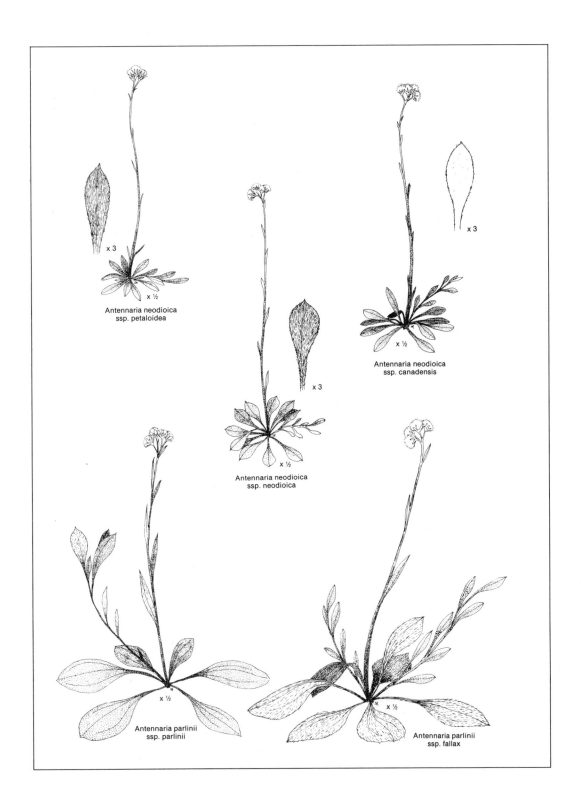

Antennaria neodioica
ssp. petaloidea

x 3

x ½

Antennaria neodioica
ssp. neodioica

x 3

x ½

Antennaria neodioica
ssp. canadensis

x 3

x ½

Antennaria parlinii
ssp. parlinii

x ½

Antennaria parlinii
ssp. fallax

x ½

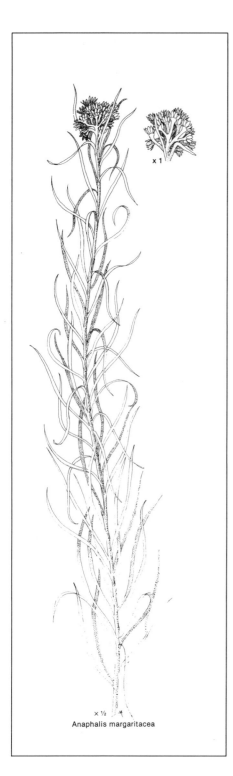

× 1

× ½

Anaphalis margaritacea

Staminate and carpellate plants about equally represented in the population; rosette leaves small, rarely over 2.5 cm long, 6 mm wide, canescent, cuneate-lanceolate to oblanceolate, acute; cauline leaves 4–8; phyllaries with erose tips, upper leaves of staminate plants with scarious tips. Rare in Ohio, known only from a single collection, but more thorough collecting will probably reveal additional sites, particularly in southeastern Ohio. Dry, open, rocky places. April–midsummer.

### 21. Anaphalis DC.   PEARLY EVERLASTING

Imperfectly dioecious, carpellate heads sometimes with a few central staminate flowers. Heads discoid, many-flowered. Phyllaries imbricate, dry, scarious, white or creamy-white. Receptacle flat or slightly convex, naked. Pappus of capillary bristles, the bristles not barbellate.

**1. Anaphalis margaritacea** (L.) Benth. & Hook.   PEARLY EVERLASTING
Plants white-woolly; stems usually unbranched; leaves lanceolate to more commonly linear, numerous, somewhat crowded, their margins revolute; inflorescence corymbose, leafy, the heads numerous; phyllaries many, obtuse or rounded. Represented in Ohio by only a few collections, the species occurs primarily in northeastern and southeastern counties, chiefly in dry, open habitats. June–Aug.

### 22. Gnaphalium L.   CUDWEED

Heads discoid. Phyllaries somewhat imbricate, dry, scarious, white or slightly colored. Receptacle flat or slightly convex, naked. Outer flowers of head carpellate, yellowish, inner flowers perfect. Pappus of capillary bristles. Achenes terete or slightly compressed. Woolly annual or perennial herbs with alternate, entire leaves.

a. Pappus bristles united in a ring at base; inflorescence narrowly spiciform and often purple-tinged . . . . . . . . . . . . . . . . . . . . . . . . . .1. G. *purpureum*
aa. Pappus bristles separate, not united at base.
  b.  Leaves decurrent onto stem; stems glandular-hairy . . . . . .2. G. *viscosum*
  bb. Leaves not decurrent onto stem; stem glandular-hairy or woolly.
    c.  Heads in glomerules subtended by overtopping leaves; stems much-branched, short, rarely to 25 cm tall . . . . . . . . . . . . . .3. G. *uliginosum*
    cc. Heads in glomerules not subtended by overtopping leaves, arranged in corymbiform or paniculate clusters; stems branched only near summit, 50–80 cm tall . . . . . . . . . . . . . . . . . . . . . . . . . . . . .4. G. *obtusifolium*

**1. Gnaphalium purpureum** L.   PURPLE CUDWEED
Stems simple or branched, usually not as woolly as G. *viscosum* and G. *uliginosum*; leaves reduced upward, lower leaves spatulate, rounded at tip, the upper

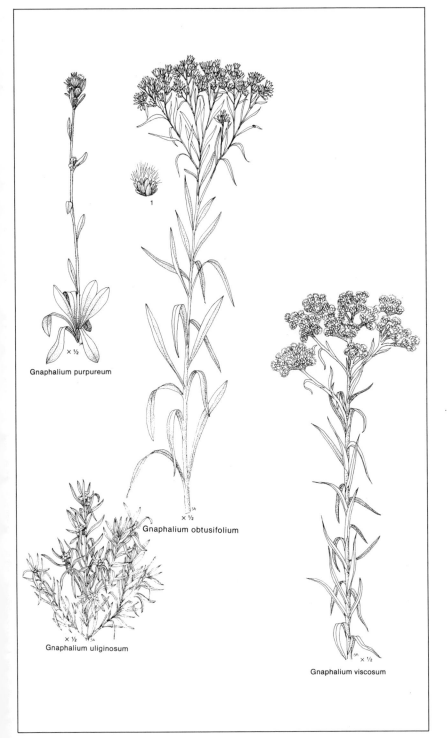

1

Gnaphalium purpureum

× ½

Gnaphalium obtusifolium

× ½

Gnaphalium uliginosum

× ½

Gnaphalium viscosum

× ½

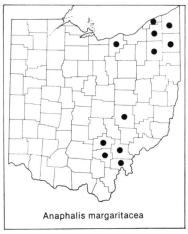

Anaphalis margaritacea

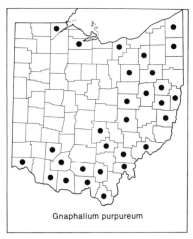

Gnaphalium purpureum

Gnaphalium viscosum

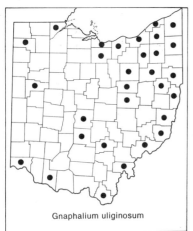

Gnaphalium uliginosum

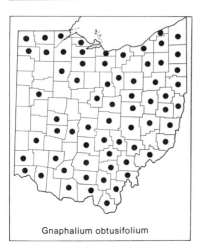

Gnaphalium obtusifolium

more linear; heads numerous, often in an interrupted spiciform arrangement, leafy bracts often present, very woolly in inflorescence; phyllaries imbricated, tawny brown or sometimes infused with purple or pink; pappus bristles united at base and falling as a ring. In various habitats, but mostly in marshes and along stream banks. May–June.

2. **Gnaphalium viscosum** H.B.K.

   *G. macounii* Greene

Similar to G. *obtusifolium*. Stems most often glandular-hairy, woolly near summit; leaves decurrent onto stem. Rare in Ohio, with records from open places in only two northern counties. Midsummer–early fall.

3. **Gnaphalium uliginosum** L.   Low Cudweed. Marsh Cudweed

Annual; stems usually branched, densely white-woolly; leaves numerous, linear to slightly oblanceolate, overtopping the inflorescence; heads numerous, in glomerules of small clusters; phyllaries small, brown or tan; pappus bristles separate, not united at base, the bristles falling separately. The purplish tinge of the phyllaries and the spiciform arrangement of the heads easily separate this species from the others. Rather common, but populations scattered, in marshes and fields. Midsummer–early fall.

4. **Gnaphalium obtusifolium** L.   Catfoot

Apparently growing as an annual; stems taller than in G. *uliginosum*, branched in the inflorescence and only sparingly white-woolly; leaves linear-lanceolate, sessile, white-woolly beneath, green and hardly pubescent above; inflorescence round-topped; phyllaries white or tawny, scarious, woolly near base; pappus bristles separate. A very common species, it occurs in a variety of habitats such as woodland borders, fields, and even more moist places. July–Oct.

## 23. Inula L.   ELECAMPANE

Heads radiate, large, to 5 cm, hemispheric. Phyllaries imbricated in several series, the outer large, pubescent, herbaceous. Disk flowers yellow, perfect. Ray flowers yellow, many, slender, carpellate. Pappus capillary, barbellate (seldom glabrous), often unequal. Receptacles flat or convex, naked. Achenes usually with 4–5 ribs or angles. Coarse perennial, hairy herbs, to 2.5 m tall. Cauline leaves alternate, cordate-clasping, the lower leaves often long-petiolate, velvety pubescent beneath, nearly glabrous or with a few spreading hairs above, the margins irregularly dentate.

1. Inula helenium L.   Elecampane

Introduced and naturalized from Europe. Cultivated and often escaped to disturbed habitats. June–Aug.

× 1

× 3

× ½

Inula helenium

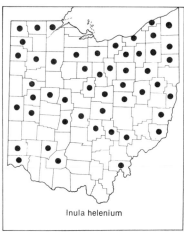

Inula helenium

# TRIBE V. HELIANTHEAE CASS.

a. Heads discoid.
  b. Flowers of the head imperfect, the outer carpellate, the inner staminate; involucre with rounded phyllaries . . . . . . . . . . . . . . . . . . . . . . .24. *Iva*
  bb. Flowers of the head imperfect, but never with both kinds of flowers in the same head.
    c. Staminate heads with involucres of united phyllaries; involucre of carpellate heads with a row of tubercules near the summit; mature heads one-seeded . . . . . . . . . . . . . . . . . . . . . . . . . . . . . . . . . . . .25. *Ambrosia*
    cc. Staminate heads with involucres of distinct phyllaries; carpellate heads of 2 flowers, maturing to become a hard, prickly, 2-seeded bur . . . . . . .
    . . . . . . . . . . . . . . . . . . . . . . . . . . . . . . . . . . . . . . . . . . . .26. *Xanthium*
aa. Heads radiate (except a few species of *Bidens*), the heads of a few species seemingly discoid, but close examination revealing short rays, often as short as 1 mm.
  b. Rays light purple to reddish-purple, or white.
    c. Rays light purple to reddish-purple. . . . . . . . . . . . . . . . .34. *Echinacea*
    cc. Rays white.
      d. Leaves opposite.
        e. Phyllaries 10–12 . . . . . . . . . . . . . . . . . . . . . . . . . . . . .32. *Eclipta*
        ee. Phyllaries 4–5 . . . . . . . . . . . . . . . . . . . . . . . . . . . . .41. *Galinsoga*
      dd. Leaves alternate.
        e. Stems winged; fruits forming in ray and disk flowers. . .37. *Verbesina*
        ee. Stems not winged; fruits forming in ray flowers only . . . . . . . . . . . .
        . . . . . . . . . . . . . . . . . . . . . . . . . . . . . . . . . . . . . . . .30. *Parthenium*
  bb. Rays yellow to orange, rarely brown or purple (whitish or whitish-yellow in one species of *Polymnia*).
    c. Stems winged . . . . . . . . . . . . . . . . . . . . . . . . . . . . . . . .37. *Verbesina*
    cc. Stems not winged.
      d. Receptacles flat or slightly convex (easiest to observe after flowers are removed from the head).
        e. Phyllaries of two kinds, the outer foliaceous.
          f. Pappus of rigid, often barbed, awns; inner phyllaries striate.
            g. Achenes terete. . . . . . . . . . . . . . . . . . . . . . . .39. *Megalodonta*
            gg. Achenes flattened or quadrangular . . . . . . . . . . . . . . . .40. *Bidens*
          ff. Pappus absent or a slight ridge, or of two awns or teeth.
            g. Rays few, about 5; leaves simple; fruits forming in ray flowers only . . . . . . . . . . . . . . . . . . . . . . . . . . . . . . . . . .29. *Chrysogonum*
            gg. Rays more than 5; leaves often compound; fruits forming in disk flowers only . . . . . . . . . . . . . . . . . . . . . . . . . . . . . .38. *Coreopsis*

ee. Phyllaries in a single series, or in several series, but all similar; pappus none or of two scales.

 f. Phyllaries in a single series; plants viscid-hairy; leaves large, lobed, with small stipule-like appendages at the base of the petiole; rays yellow or whitish . . . . . . . . . . . . . . . . . . . . . . . . . . . . .27. *Polymnia*

 ff. Phyllaries in several series and similar.

  g. Phyllaries ovate, blunt, or if acute the leaf blades perfoliate or pinnatifid; achenes flat, winged, forming in ray flowers only . . . . . . . . . . . . . . . . . . . . . . . . . . . . . . . . . . . . . . . . . . .28. *Silphium*

  gg. Phyllaries acute, much longer than wide; achenes thick, wingless, forming in disk flowers only . . . . . . . . . . . . . . . .36. *Helianthus*

dd. Receptacles markedly convex to conic or columnar.

 e. Leaves opposite . . . . . . . . . . . . . . . . . . . . . . . . . . . . . . .31. *Heliopsis*

 ee. Leaves alternate.

  f. Achenes 4-sided . . . . . . . . . . . . . . . . . . . . . . . . . . . . .33. *Rudbeckia*

  ff. Achenes flattened or 3-sided. . . . . . . . . . . . . . . . . . . . . .35. *Ratibida*

## 24. Iva L.  MARSH-ELDER

Heads discoid, small, greenish. Central flowers of head staminate, marginal flowers few and carpellate, corollas nearly lacking. Phyllaries few, imbricated in 1–3 series. Receptacle small, chaffy. Anthers weakly united. Achenes essentially glabrous, obovoid. Pappus none. Annual herbs. Leaves mostly opposite. The genus was revised by Jackson (1960).

a. Phyllaries 5 or fewer; heads sessile on spike-like branches, each head subtended by a single lanceolate bract . . . . . . . . . . . . . . . . . . . . .1. *I. annua*

aa. Phyllaries 10; heads on panicle-like branches, heads without subtending bracts . . . . . . . . . . . . . . . . . . . . . . . . . . . . . . . . . . . . .2. *I. xanthifolia*

1. IVA ANNUA L.  SUMPWEED

 *Iva caudata* Small

 *Iva ciliata* Willd.

Stems glabrous below, hirsute above; leaves opposite below, alternate above, lanceolate to ovate-lanceolate, obscurely serrate, pubescent on both surfaces; heads arranged in several spiciform branches, the heads sessile in axils of linear bracts; phyllaries long, hirsute; carpellate flowers with persistent corolla; achenes obovate, glabrous. Weedy annual introduced from more midwestern and western parts of North America. Rare. Late summer–Oct.

2. IVA XANTHIFOLIA Nutt.  MARSH-ELDER

Stems glabrous below, viscid-pubescent above, especially in inflorescence branches; leaves opposite below, alternate above, broadly ovate, pubescent, more regularly serrate than in *I. annua*, long-petiolate; inflorescence more

Iva annua

Iva xanthifolia

paniculate than spicate; heads not subtended by leafy bracts; phyllaries glabrous to pubescent; carpellate flowers without persistent corollas. Weedy annual introduced from further west. Moist waste places. Rare. Aug.–Oct.

× ½

Iva annua

×½
Iva xanthifolia

## 25. Ambrosia L. RAGWEED

Flowers imperfect, plants monoecious. Staminate heads small, discoid, in spikes or racemes. Phyllaries united in a 5–12-lobed involucre, corolla 5-lobed. Carpellate heads borne below the staminate in axils of bracts, flowers without corollas, solitary in a closed, globose, ovoid, or turbinate, spiny or tuberculed involucre. Pappus none. Annuals or perennials. Leaves lobed or dissected.

*Ambrosia* is represented in Ohio by four species, three of which are troublesome weeds of cultivated fields and waste places. The wind-borne pollen is a significant cause of hay fever.

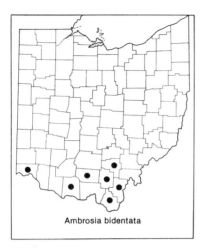
Ambrosia bidentata

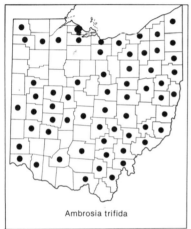
Ambrosia trifida

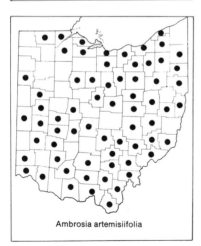
Ambrosia artemisiifolia

a. Staminate heads sessile, involucre zygomorphic; leaves entire or hastate
......................................................1. *A. bidentata*

aa. Staminate heads pedunculate, involucre not or only slightly zygomorphic.

  b. Leaves opposite, palmately 3–5-lobed or some entire; plants tall and coarse ..........................................2. *A. trifida*

  bb. Leaves pinnatifid or more often bipinnatifid, opposite or alternate.

    c. Leaves pinnatifid or bipinnatifid; involucre with 5–8 sharp spines; plants annual, lacking rhizomes ...............3. *A. artemisiifolia*

    cc. Leaves once pinnatifid, hoary-pubescent; involucre with beak and 1–4 short tubercules; plants with creeping rhizomes .....4. *A. psilostachya*

1. **Ambrosia bidentata** Michx.

Annual, usually well-branched, stems spreading-hirsute above; leaves sessile, opposite below, alternate above, upper leaves lanceolate, entire, the middle and lower hastate by virtue of large lateral teeth near base of blade, most often hirsute but occasionally glabrous; staminate spikes solitary, heads crowded, sessile, the turbinate involucre ovoid, 4-angled or -ribbed, the angles extended into 4 stout spines. Open waste places, probably more widely distributed than collection records indicate. July–Oct.

2. **Ambrosia trifida** L.    HORSEWEED. GIANT RAGWEED

Very tall, stout annual, up to 4 m tall, stems rough-hairy; leaves opposite, petiolate, nearly always 3-lobed, lobes ovate-lanceolate, serrate, scabrous on both surfaces; staminate spikes solitary, heads on slender peduncles, crowded; carpellate involucre several-ribbed, each rib ending in a blunt tubercle or spine. A common weedy species of flood plains and waste places, probably occurring in every Ohio county. July–Oct.

Plants with entire leaves have been named forma *integrifolia* (Muhl.) Fern., but I believe them to be merely depauperate forms of the species, and of no taxonomic significance.

3. **Ambrosia artemisiifolia** L.    COMMON RAGWEED. WORMWOOD.

HOGWEED

  *A. elatior* L.

Annual, stems hairy or glabrous, branched, to 2 m tall; leaves opposite, or sometimes alternate above, petiolate, pinnatifid or more commonly bipinnatifid; staminate heads short-pedunculate; involucre of carpellate heads with central tubercle longer than the 4–7 sharp lateral ones. A rather variable species common to cultivated fields and waste places, probably occurring in all Ohio counties. Aug.–Oct.

Natural hybridization between *A. artemisiifolia* and *A. trifida* was reported by Wagner (1958) in a well-documented study of a population in Wood County. The hybrid, *A.* x *helenae* Rouleau, is probably common, but often escapes notice.

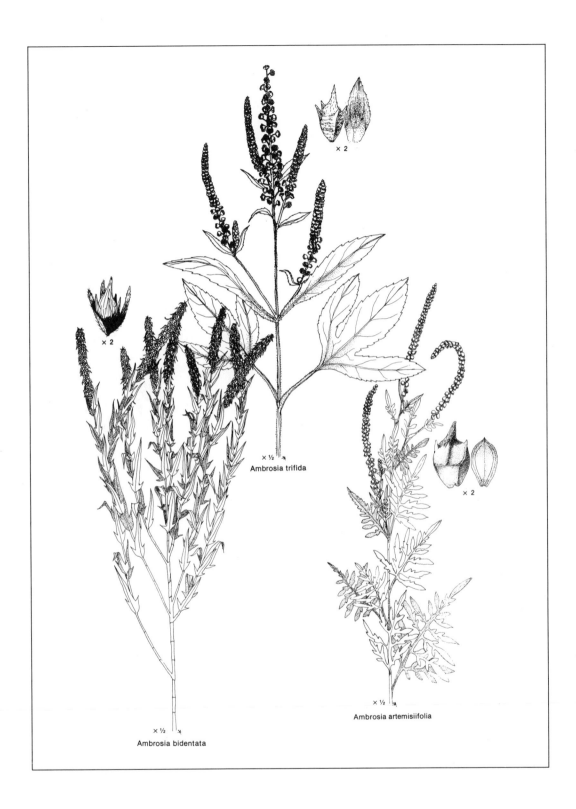

× 2

× 2

× 2

× ½
Ambrosia trifida

× ½
Ambrosia artemisiifolia

× ½
Ambrosia bidentata

× ½

Ambrosia psilostachya

× ½

Xanthium spinosum

4. AMBROSIA PSILOSTACHYA DC.    WESTERN RAGWEED
  *A. coronopifolia* T. & G.

Perennial with creeping rhizomes; stems to 1 m tall, hoary, harsh to the touch, the pubescence stiff and short; leaves once-pinnatifid, short-petiolate to sessile; staminate involucres minutely scabrous; carpellate involucres without tubercles or with very short ones. Dry, sandy habitats. Rare in Ohio; adventive from states to the west. July–Oct.

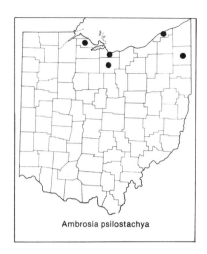

Ambrosia psilostachya

## 26. Xanthium L.   COCKLEBUR

Flowers imperfect, plants monoecious. Staminate heads discoid, small, many-flowered, in short spikes or racemes, involucre of separate phyllaries. Receptacle chaffy, corollas tubular, anthers separate. Carpellate heads with united phyllaries covered with hooked or straight prickles, 2-flowered, corolla none. Fruit a prickly bur, a single achene in each of the 2 locules. Pappus none. Annual herbs. Leaves opposite.

Cosmopolitan weeds widely distributed in waste places, in cultivated fields, on flood plains, and on beaches.

a.   Leaves lanceolate, tapering at both ends, with 3-branched axillary spines
     present . . . . . . . . . . . . . . . . . . . . . . . . . . . . . . . . . . . . . . . . . . .1. *X. spinosum*
aa.  Leaves broadly ovate, axillary spines absent. . . . . . . . . . .2. *X. strumarium*

1. XANTHIUM SPINOSUM L.    SPINY COCKLEBUR

Stems branched, strigose or somewhat puberulent; leaves lanceolate, tapering at both ends, mostly entire, short-petiolate, glabrous to slightly strigose above, sericeous beneath, the axils armed with 3-branched spines; burs small, less than 1 cm, spines decidedly hooked. A rare plant of disturbed habitats, without doubt a transient from the southern and western states. July–Oct.

Xanthium spinosum

2. **Xanthium strumarium** L.    COMMON COCKLEBUR
  *X. americanum* Walt.
  *X. chinense* Mill.
  *X. globosum* Shull
  *X. inflexum* Mackenzie & Bush
  *X. italicum* Moretti
  *X. pensylvanicum* Wallr.
  *X. speciosum* Kearney

Stems branched, hairy to nearly glabrous, often mottled with irregular dark areas; leaves broadly ovate, often cordate at the base, often shallowly lobed, long-petiolate; heads several on short axillary branches; burs (fruits) armed with strong hooked spines, the spines glabrous, puberulent, or hairy. A very weedy species of cultivated fields, flood plains, and waste places. Young plants are extremely poisonous. Midsummer–Oct.

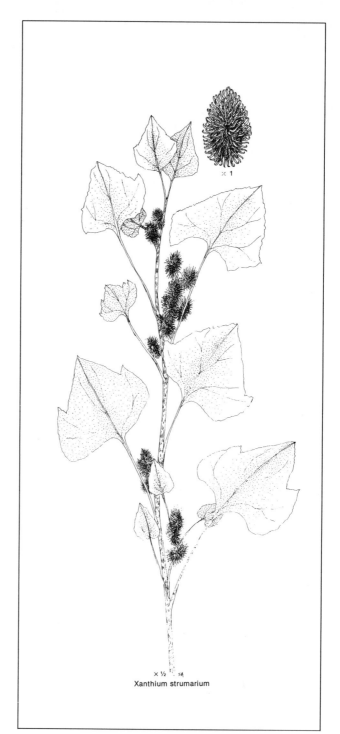

Xanthium strumarium
var. canadense

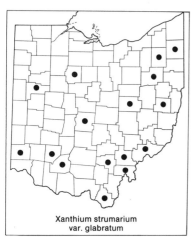

Xanthium strumarium
var. glabratum

x 1

x ½

Xanthium strumarium

The species consists of three varieties, two of which are found in Ohio, namely var. *canadense* (Mill.) T. & G. and var. *glabratum* (DC.) Cronq. The bur of var. *canadense* (illustrated) is pubescent on and among the spines, whereas that of var. *glabratum* is puberulent or glabrous. The typical var. *strumarium* is not found in Ohio.

## 27. Polymnia L. LEAFCUP

Heads radiate or discoid. Phyllaries 5 in a single series. Receptacle flat, chaffy, the outer pales large and partly embracing the achenes. Ray flowers white or yellow, sometimes small and lacking the expanded (limb) portion of the corolla, resulting in the discoid condition, otherwise carpellate and seed-forming. Disk flowers staminate only. Achenes glabrous or minutely hairy. Pappus absent. Rather tall perennial plants with mostly opposite, viscid-pubescent or rough-hairy leaves and stems. For the basic taxonomy and monograph, see Wells (1965).

a. Rays about 5, whitish or whitish-yellow, very short to about 1 cm long; achenes definitely 3-ribbed; petioles not winged . . . . . . . .1. *P. canadensis*

aa. Rays 10–15, yellow, 1–3 cm long or rarely none; achenes merely striate; petioles winged to base . . . . . . . . . . . . . . . . . . . . . . . . . . .2. *P. uvedalia*

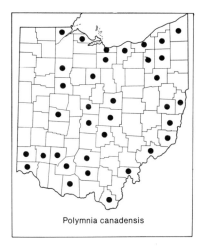

Polymnia canadensis

### 1. **Polymnia canadensis** L. LEAFCUP

Coarse perennial to 2 m tall; stems glabrous below to viscid-villous or clammy-hairy above, especially in the inflorescence; lower leaves deeply pinnatifid, the upper triangular-ovate and 3–4-lobed, petiolate; heads small, rarely over 1 cm wide, in cymose clusters; phyllaries lanceolate, often shorter than the pales that subtend the ray achenes; ray flowers carpellate, small and tubular or, if expanded, the limb very short; achenes often large, to 4.5 mm, 3-ribbed or -angled, smooth. Common in moist woodlands, especially on calcareous soils. June–Oct.

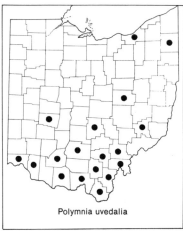

Polymnia uvedalia

### 2. **Polymnia uvedalia** (L.) L. BEAR'S-FOOT. YELLOW LEAFCUP

Coarse perennial to 2 m tall; stems glandular-pubescent, especially on upper parts and in inflorescence, otherwise essentially glabrous; leaves broadly ovate, subpalmately lobed, hispid to subglabrous above, often glandular-hispid beneath, petioles winged; heads in a more open arrangement than those of *P. canadensis*; phyllaries lance-ovate, herbaceous; ray corollas yellow, to 3 cm long, sometimes reduced; achenes merely striate. Mostly in moist woods and openings of the south-central and southeastern parts of the state. Apparently this species occurs more often in acid soils than does *P. canadensis*. July–Sept.

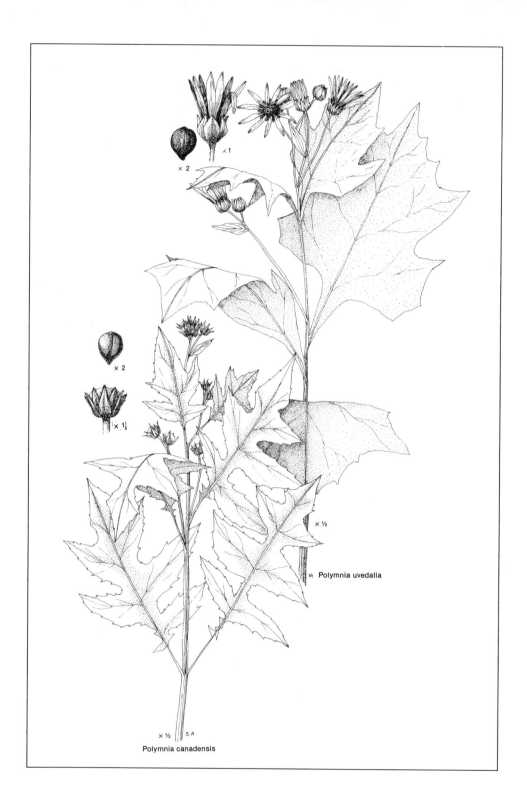

× 2

× 1

× 2

× 1

× ½

*Polymnia uvedalia*

× ½   5 A

*Polymnia canadensis*

## 28. Silphium L.  ROSIN-WEED

Heads radiate, phyllaries in 2–3 series, herbaceous, unequal. Receptacle flat, chaffy throughout. Disk flowers perfect, not seed-forming. Ray flowers yellow, imbricate in 2–3 series, carpellate, seed-forming. Pappus none or of 2 awns that are a continuation of the achene wings. Achenes broadly flattened, winged along the margins. Large, coarse, perennial herbs with alternate, opposite, whorled, or basal leaves. See Fisher (1966) for study of the genus in Ohio.

About 20 species native to central, south, and southeastern United States. Four species occur in Ohio, although one, *S. laciniatum*, is rare and known from only one collection site.

a. Stems square; leaves connate-perfoliate . . . . . . . . . . . . . .1. *S. perfoliatum*
aa. Stems round; leaves not connate-perfoliate.
  b. Stems naked except for scattered bracts; principal leaves basal . . . . . . . .
  . . . . . . . . . . . . . . . . . . . . . . . . . . . . . . . . . . . . .2. *S. terebinthinaceum*
  bb. Stems leafy.
    c. Leaves pinnatifid or bipinnatifid, alternate . . . . . . . . . .3. *S. laciniatum*
    cc. Leaf margins serrate or occasionally entire, ovate-lanceolate, opposite, alternate, or sometimes whorled . . . . . . . . . . . . . . . . . .4. *S. trifoliatum*

### 1. **Silphium perfoliatum** L.  CUP-PLANT

Stems to 3 m tall, square, glabrous; leaves connate-perfoliate, ovate, coarsely toothed, scabrous above and beneath, leathery; heads numerous, inflorescence somewhat corymbose; phyllaries imbricate, the outer ovate, the inner more narrow, glabrous except for ciliate margins; ray flowers light yellow; achenes obovate, winged, sinus broad at apex. A common species of moist woodland margins and flood plains. July–Sept.

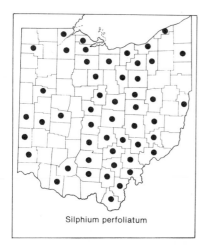

Silphium perfoliatum

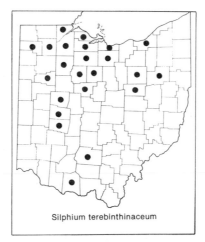

Silphium terebinthinaceum

x ½

x 2

x 2

Silphium laciniatum
× ½

× ½

Silphium perfoliatum

## 2. **Silphium terebinthinaceum** Jacq.  PRAIRIE-DOCK

Stems to 3 m tall, glabrous, naked except for small leafy bracts; taproot thick and woody; leaves large, the basal broadly ovate to ovate-oblong, cordate at base, sharply toothed, generally scabrous; heads several, large, in a corymbose arrangement or in large open panicles; phyllaries imbricate, the outer broadly ovate-elliptic, the inner more slender, glabrous; achenes broadly winged, apical notch broad and shallow, hardly toothed. Prairie habitats. Common, especially in the northern and northwestern parts of Ohio. Aug.–early Oct.

*Silphium terebinthinaceum* hybridizes occasionally with *S. laciniatum*, and the hybrids are often erroneously identified as *S. pinnatifidum* Ell., a more southern species (Fisher, 1959). Plants from Jefferson Township in Adams County with slightly smaller and glabrous leaves can be identified as var. *lucy-brauniae* Steyerm. (see Steyermark, 1951).

## 3. **Silphium laciniatum** L.  COMPASS-PLANT

Stems to 3 m tall, coarse-hispid, from woody taproot; leaves alternate, pinnatifid to bipinnatifid, hirsute along midrib and veins beneath, basal or lower leaves large, petioles clasping at base, leaves gradually reduced upward, the upper smaller and entire; heads few, on short peduncles or sessile; inflorescence racemose; phyllaries ovate, tapering to long, stiff points, hispid-hirsute and often somewhat reflexed; achenes broadly winged, apical notch deep. Prairies and open places. July–Sept.

Rare in Ohio (see monograph by Anderson, 1968). Presently known from but a single living population in Lawrence County (Fisher, 1959). For other papers dealing with the ecology and distribution of this species, see Cusick (1981) and Auffenorde and Wistendahl (1983).

Silphium laciniatum

## 4. **Silphium trifoliatum** L.  WHORLED ROSIN-WEED

Stems to 2 m tall, glabrous or sometimes glaucous; leaves opposite, alternate, or often whorled in 3s or even 4s, narrowed to a short petiole, ovate to ovate-lanceolate, toothed to nearly entire, mostly scabrous above, glabrous to hirsute on main veins beneath; heads in an open corymbose arrangement; phyllaries imbricate, glabrous on back, ciliolate on margins; ray flowers 8–15, lemon-yellow, attractive; achenes obovate to orbicular, wing well-developed, toothed at apex. For a detailed monograph of this species, see Weber (1968). Common at woodland borders and in prairie habitats. July–Sept.

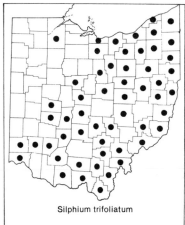

Silphium trifoliatum

×1

×2

×1

×2

Silphium trifoliatum

×½

Silphium terebinthinaceum

×½

## 29. Chrysogonum L.

Heads radiate, few to solitary on slender branches. Phyllaries 4–5 in 2 series, the outer herbaceous and longer than the disk and alternating with the ray flowers, the inner smaller and partly enclosing the achenes. Receptacle flat, chaffy. Disk flowers perfect, not seed-forming. Ray flowers usually 5, rather broad, yellow, carpellate, seed-forming. Achenes flattened, wingless. Pappus none or a mere crown of 2–3 teeth. Leaves opposite, ovate, long-petiolate, hairy, dentate.

### 1. **Chrysogonum virginianum** L.

Rare woodland species occurring in only a few southeastern counties, although common in southeastern United States. In Athens County it is local but occurs in rather large populations. Early spring–July.

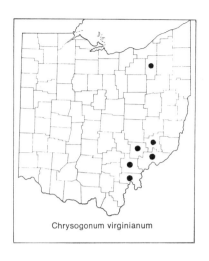

Chrysogonum virginianum

× 1

× ½

Chrysogonum virginianum

# 30. Parthenium L.

Heads radiate, in a small corymbose arrangement. Phyllaries 2–4, somewhat imbricate, dry. Receptacle, conic or convex, chaffy. Disk flowers perfect, not seed-forming. Ray flowers few, about 5, short, white or cream, somewhat inconspicuous, carpellate, seed-forming. Pappus essentially absent or of 2–3 scales or awns. Perennial or annual herbs, with alternate entire or pinnatifid leaves.

a.   Plants annual; leaves pinnatifid or bipinnatifid . . . . . . .1. *P. hysterophorus*
aa. Plants perennial; leaves crenate, serrate, or dentate . . . .2. *P. integrifolium*

1. Parthenium hysterophorus L.

Annual, to 1 m tall, abundantly branched, glabrous to pubescent; leaves pinnatifid to bipinnatifid, short-pubescent on both surfaces; heads on slender peduncles arranged in a loose corymbiform cyme; rays small, cream or whitish; achenes ovoid, black; pappus of 2 small scales or teeth. A plant of dry, waste places; native to tropical America. July–Oct.

2. **Parthenium integrifolium** L.   Wild Quinine

Perennial from thickened roots; stem glabrous or only slightly pubescent; basal leaves with long petioles, blades thick, harsh-pubescent above, glabrous to minutely pubescent beneath, ovate to oblong, margins crenate to serrate, cauline leaves few, the upper sessile, dentate and scabrous; inflorescence corymbose; phyllaries ovate, canescent; pappus of two small teeth or scales. For the basic taxonomic treatment see Rollins (1950). Open woodlands and prairies. July–Oct.

I have seen only one Ohio specimen, collected from Cuyahoga County, but Rollins (1950) and Mears (1975) cite specimens from Geauga County. Rare, possibly extirpated.

Parthenium hysterophorus

Parthenium integrifolium

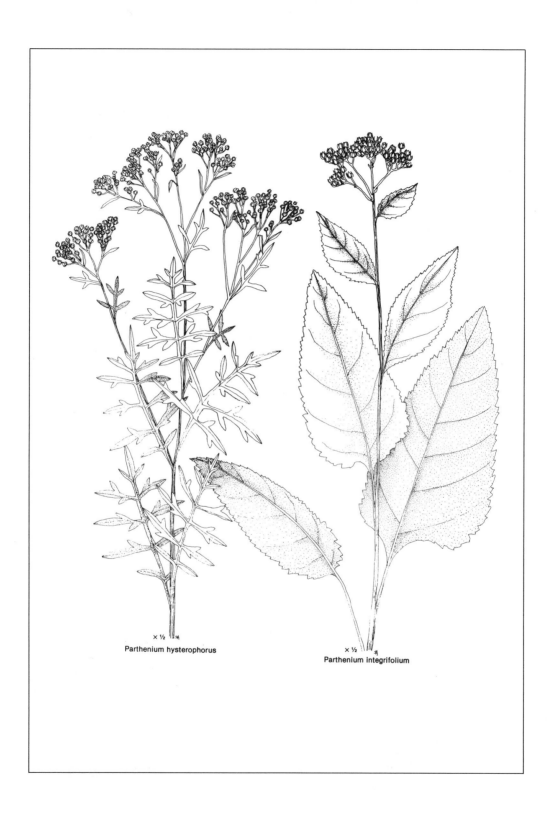

× ½

Parthenium hysterophorus

× ½

Parthenium integrifolium

## 31. Heliopsis Pers. OX-EYE

Heads radiate. Phyllaries in 1 or more often in 2 series, subequal, the outer leafy. Receptacle conic, chaffy throughout, the pales embracing the achenes. Disk flowers perfect and seed-forming. Ray flowers yellow, carpellate and seed-forming. Pappus none, or a mere crown, or of a few minute and weak teeth. Achenes of disk flowers quadrangular, those of the rays 3-sided. Perennial herbs with opposite leaves. The treatment of the genus is based on that of Fisher (1957).

### 1. **Heliopsis helianthoides** (L.) Sweet   SMOOTH OX-EYE

Well-represented in the Ohio flora and, flowering in June, the vanguard to a host of yellow-flowered composites. Found primarily in woodlands, flood plains, and prairies. June–Sept.

The ssp. *helianthoides* is most common and grades into ssp. *occidentalis* Fisher in northwestern and north-central Ohio. (Fisher, 1958) The ssp. *helianthoides* can be recognized by its nearly smooth stems, ovate leaves on long petioles, and small heads on short peduncles. The ssp. *occidentalis* has very scabrous leaves on short petioles and generally larger heads on long peduncles. The latter also grows in more open prairie habitats.

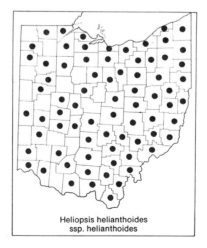

Heliopsis helianthoides
ssp. helianthoides

Heliopsis helianthoides
ssp. occidentalis

x 1

x ½

Heliopsis helianthoides
ssp. occidentalis

Heliopsis helianthoides
ssp. helianthoides

## 32. Eclipta L.

Heads small, radiate. Phyllaries in 1–2 series, somewhat subequal, pubescent on back. Receptacle mostly flat or slightly convex, chaffy, the pales slender or even bristle-like, often absent in center. Disk flowers perfect, seed-forming, corollas 4- or 5-toothed. Ray flowers white, carpellate, usually seed-forming, small, barely exceeding the disk. Pappus absent, or a slight crown, or of 2 short awn-like teeth. Achenes 3–4-angled or slightly compressed, transversely rugose. Annual herbs with weak stems, often spreading and rooting at nodes. Leaves opposite, lanceolate or linear, slightly serrate or entire, sessile or very short-petiolate.

Represented in Ohio by a single rather widespread species.

### 1. **Eclipta alba** (L.) Hassk.

A common weedy species of flood plains and moist places around ponds and lakes. Native to the New World and widespread in the tropics and warm-temperate climates. Aug.–Oct.

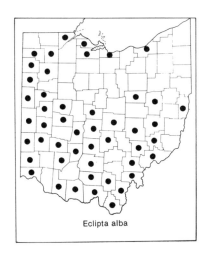

Eclipta alba

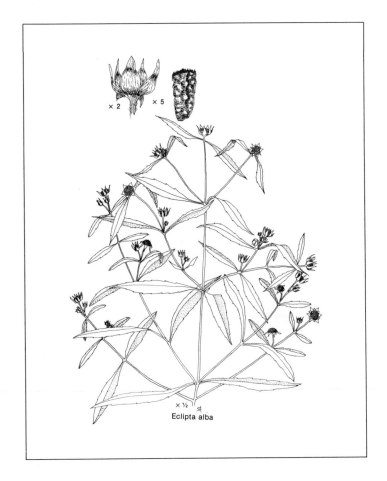

× ½
Eclipta alba

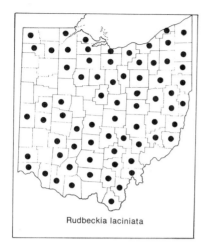

Rudbeckia laciniata

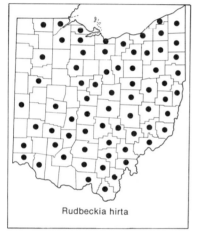

Rudbeckia hirta

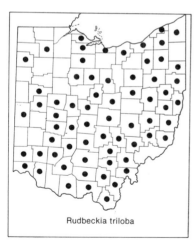

Rudbeckia triloba

## 33. Rudbeckia L. CONEFLOWER

Heads radiate, large, usually on long peduncles. Phyllaries in 2–3 series, subequal, spreading and reflexed, herbaceous. Receptacle somewhat enlarged, convex to conic, chaffy, pales partly embracing achenes. Disk flowers perfect and seed-forming. Ray flowers neutral, yellow, orange, or purplish. Pappus none or a slight crown. Achenes quadrangular. Perennial herbs with alternate, unlobed to 3-lobed or pinnatifid leaves.

Two specimens deposited at The Ohio State University Herbarium, collected in Fairfield and Wayne Counties, appear to fit the description of *R. subtomentosa* Pursh. This species is not included here because of its dubious status in the Ohio flora.

a. Disk corollas gray, green, or yellow; at least some lower leaves cleft or pinnatifid . . . . . . . . . . . . . . . . . . . . . . . . . . . . . . . . . . . . . . .1. *R. laciniata*
aa. Disk corollas purple or brown.
  b. Pappus none; style branches elongate and narrowly subulate; pales with small bristle-tips . . . . . . . . . . . . . . . . . . . . . . . . . . . . . . . . . .2. *R. hirta*
  bb. Pappus an obvious crown; style branches short, blunt.
    c. Pales glabrous, awn-pointed; leaves mostly 3-lobed . . . . . . .3. *R. triloba*
    cc. Pales glabrous or pubescent, not terminating in a short awn, often ciliate; leaves unlobed . . . . . . . . . . . . . . . . . . . . . . . . . . . . . .4. *R. fulgida*

### 1. Rudbeckia laciniata L.

Stems tall, glabrous, often glaucous; leaves coarsely toothed to laciniate, usually pinnatifid or 3-lobed, usually glabrous above and beneath; heads rather large, yellowish, grayish, or greenish, globose or hemispheric; phyllaries spreading, smooth, usually reflexed, leaf-like; ray flowers yellow, drooping; pales blunt, canescent at tip; achenes 4-angled. Locally abundant, most often along streams and flood plains. July–late fall.

The cultivated variety *hortensa* Bailey, often called GOLDEN-GLOW in seed catalogs, may be found as an escape in fields and along fence rows. It is easily distinguished by having about 2–3 times as many ray flowers as the typical variety.

### 2. Rudbeckia hirta L. BLACK-EYED SUSAN

Stems usually branched and hirsute throughout, plant functioning as a biennial; leaves often entire but usually dentate, the lower long-petiolate, the mid

× 2

× 1

× 2

× 1

× ½

Rudbeckia laciniata

× ½

Rudbeckia hirta

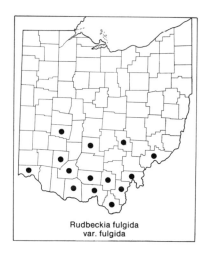

Rudbeckia fulgida
var. fulgida

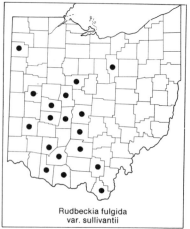

Rudbeckia fulgida
var. sullivantii

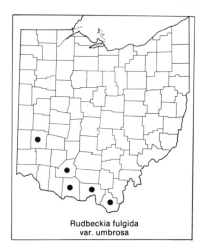

Rudbeckia fulgida
var. umbrosa

and upper mostly sessile, oblong to narrowly ovate; heads borne singly in poor habitats, but usually many, hemispheric, dark purple to deep brown, peduncles long; phyllaries hirsute, subequal, reflexed; ray flowers yellow or orange-yellow; pales acute, hispid, hispid-ciliolate on margins; achenes 4-angled. A very widespread species occurring in a wide variety of habitats. June–Oct.

Of the two varieties often recognized, only var. *pulcherrima* Farw. (*R. serotina* Nutt.) occurs in Ohio.

### 3. **Rudbeckia triloba** L.

Stems branched, hirsute, perennial; leaves subentire to nearly evenly dentate, sparingly hirsute, cauline leaves ovate-lanceolate, sessile, basal leaves broadly ovate, long-petiolate, often 3-lobed or trifoliolate; heads many, dark purple, hemispheric; phyllaries subequal, spreading, reflexed, strigose, ciliolate; ray flowers yellow, yellowish-orange or with an orange base; pales awn-tipped; achenes 4-angled. Open places and thickets. July–late fall.

This species is widespread in Ohio and often confused with *R. fulgida* Ait., but the awn tip or cusp of the pales clearly distinguishes it from the other species of this genus. July–late fall.

### 4. **Rudbeckia fulgida** Ait.   ORANGE CONEFLOWER

Stems perennial, branched, spreading-hirsute to nearly glabrous; leaves quite variable, linear-lanceolate to ovate and even cordate, coarsely dentate to entire, sessile to long-petiolate; heads usually several, usually on long peduncles, disks dark brown to purple; rays yellow or yellow and orange; pales obtuse or acute, glabrous or pubescent or ciliolate on margins; achenes 4-angled. Moist or dry, shaded or open habitats. July–Oct.

This variable species is not particularly abundant in Ohio, and is often confused with *R. triloba* L. It would be quite difficult not to recognize three varieties in Ohio, even though at times they are quite confusing. In their "pure" form, each variety is easily recognized although var. *sullivantii* (Boynt. & Beadle) Cronq. and var. *fulgida* appear to intergrade. Variety *umbrosa* (Boynt. & Beadle) Cronq., the rarest, is restricted chiefly to the more southern counties.

a. Basal leaves narrowly elliptic or lanceolate, cauline leaves linear to lanceolate, mostly sessile, entire or merely denticulate; pales nearly always glabrous and eciliate; rays short, 1.5–3.0 cm long . . . . . . . . . . . . .var. *fulgida*
   *R. tenax* Boynt. & Beadle, *R. acuminata* Boynt. & Beadle

aa. Basal leaves mostly ovate or subcordate, long-petiolate, usually dentate, sometimes entire; pales ciliate or mostly so.

×3

×½

×½

Rudbeckia triloba

Rudbeckia fulgida
var. fulgida

Rudbeckia fulgida
var. sullivantii

Rudbeckia fulgida
var. umbrosa

b. Leaf blades coarsely dentate; leaves short-petiolate, sparsely hirsute; ray flowers 2.5–4.0 cm long . . . . . . . . . . . . . . . . . . . . . . . . . . .var. *sullivantii*

*R. sullivantii* Boynt. & Beadle, *R. speciosa* Wenderoth, *R. speciosa* var. *sullivantii* (Boynt. & Beadle) Robinson

bb. Leaf blades nearly entire to finely dentate, ovate to subcordate at base; leaves with winged, or slightly winged, villous-hirsute petiole; ray flowers 1.5–3.0 cm long . . . . . . . . . . . . . . . . . . . . . . . . . . . . . . .var. *umbrosa*

*R. umbrosa* Boynt. & Beadle

x 2

x ½

Echinacea purpurea

### 34. Echinacea Moench CONEFLOWER

Heads radiate, solitary or very few. Phyllaries in 2–4 series, somewhat imbricated, reflexed, glabrous or nearly so. Receptacle conic, chaffy, the pales slightly enclosing achene and with slender stiff tips exceeding the disk corollas. Disk flowers perfect, purple, seed-forming; pappus a short, slightly toothed crown. Achenes 4-sided, not formed in the drooping, reddish-purple to light purple ray flowers. Perennial herbs. Leaves alternate, pubescent, broadly ovate, slightly toothed, the petiole sometimes winged.

Represented in Ohio by a single species.

### 1. **Echinacea purpurea** (L.) Moench PURPLE CONEFLOWER

A beautiful species of prairies and thin woodlands. June–Oct.

Two varieties are recognized in eastern United States, but only var. *purpurea* occurs in Ohio. Several cultivars, distinguished primarily on the basis of ray color, are sold in garden stores and from seed catalogs. For further information on the genus, see McGregor (1968).

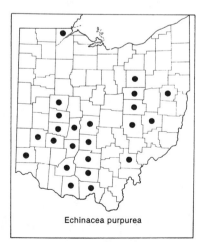

Echinacea purpurea

## 35. Ratibida Raf. PRAIRIE CONEFLOWER

Heads radiate, on long peduncles. Phyllaries uniseriate, herbaceous, lance-linear. Receptacle columnar, chaffy throughout, pales weakly embracing the achenes. Disk flowers perfect, seed-forming. Ray flowers neutral, yellow. Pappus absent or a crown of 1–2 weak teeth. Achenes flat to quadrangular. Perennial herbs of prairies, prairie remnants, woodlands, and waste places with alternate, pinnatifid leaves. See Richards (1968) for basic taxonomic treatment.

a. Disks cylindric, at least 2.5–5.0 times as high as thick; pappus of 1–2 weak teeth . . . . . . . . . . . . . . . . . . . . . . . . . . . . . . . . .1. *R. columnifera*
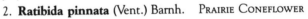
aa. Disks ellipsoid-cylindric, only slightly higher than thick; pappus none . . . . . . . . . . . . . . . . . . . . . . . . . . . . . . . . . . . . . . . .2. *R. pinnata*

### 1. RATIBIDA COLUMNIFERA (Nutt.) Woot. & Standl.

Stems from stout caudex, usually branching from near base, to 1 m or more tall; leaves rough-hirsute, pinnatifid, segments linear to lanceolate, entire; heads several, peduncles long; disk cylindric, at least 1.5 cm high, usually higher; ray flowers showy, yellow, to 3 cm long, mostly drooping or reflexed; achenes slightly winged; pappus of 1–2 unequal awns or teeth.

A western species of prairies and waste places. Probably nonindigenous in Ohio, where it has not been collected since 1899 (a Franklin County specimen at The Ohio State University Herbarium). (Not illustrated or mapped) July–Oct.

### 2. Ratibida pinnata (Vent.) Barnh. PRAIRIE CONEFLOWER

Stems from stout rhizomes; stems to 1 m tall, usually with fewer branches than *R. columnifera*; leaves hirsute to nearly scabrous, pinnatifid, segments lanceolate, coarsely toothed or more often entire; heads usually several, disk shorter than rays; rays pale yellow, spreading but more often drooping; achenes smooth; pappus none. Prairies and prairie remnants, sometimes woodland borders and railroad rights of way. Common. July–Oct.

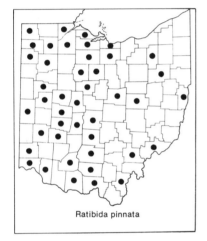

Ratibida pinnata

## 36. Helianthus L. SUNFLOWER

Heads radiate. Phyllaries imbricate in several series, herbaceous. Receptacle flat to low-conic, chaffy, the pales embracing the achenes. Disk flowers perfect, seed-forming. Ray flowers yellow, not seed-forming. Pappus of 2, thin, chaff-like awns or scales. Achenes slightly compressed. Annual or perennial herbs. Leaves simple, mostly opposite, but often alternate above. See studies by Watson (1929), Beatley (1963), Heiser and Smith (1964), and especially Heiser et al. (1969).

a. Plants annual.
  b. Phyllaries over 4 mm wide; pales short-hairy at tip . . . . . .1. *H. annuus*
  bb. Phyllaries less than 4 mm wide; pales with conspicuous white hairs at tip; anthers purple. . . . . . . . . . . . . . . . . . . . . . . . . . . . . .2. *H. petiolaris*
aa. Plants perennial.

× 1

× ½

Ratibida pinnata

b. Plants with much-reduced upper leaves, scapose to subscapose, lower leaves long-tapering to base . . . . . . . . . . . . . . . . . . . . .3. *H. occidentalis*

bb. Plants with well-developed middle and upper leaves, the leaves only somewhat reduced upward; plants not scapose.

  c. Main leaves of the stem linear to linear-oblong, to 1.5 cm wide.

    d. Corolla lobes of the disk flowers yellow; leaves conduplicate (folded lengthwise) . . . . . . . . . . . . . . . . . . . . . . . . . . . . . .4. *H. maximilianii*

    dd. Corolla lobes purple or reddish-brown; leaves rarely 5.0 mm wide, not conduplicate . . . . . . . . . . . . . . . . . . . . . . . . . . . . .5. *H. angustifolius*

  cc. Main leaves of the stem ovate to narrowly lanceolate, wider than 1.5 cm.

    d. Leaves mostly alternate.

      e. Stems glabrous; leaves mostly alternate, shallowly serrate, rarely subentire . . . . . . . . . . . . . . . . . . . . . . . . . . . . . . . . . .6. *H. grosseserratus*

      ee. Stems pubescent, scabrous-hispidulous or hirsute.

        f. Leaves usually conduplicate (folded), entire or nearly so; inflorescence often spiciform; phyllaries long-attenuate . . . . . . . . . . . . . . . . .
. . . . . . . . . . . . . . . . . . . . . . . . . . . . . . . . . . . . . . . .4. *H. maximilianii*

        ff. Leaves flat, not conduplicate, subsessile or often with short, ciliate petioles; inflorescence usually paniculate; phyllaries ciliate, rarely long and attenuate . . . . . . . . . . . . . . . . . . . . . . . . .7. *H. giganteus*

    dd. Leaves, at least the lower, mostly opposite.

      e. Phyllaries tightly appressed, without long-tapering tips; leaves mostly basal . . . . . . . . . . . . . . . . . . . . . . . . . . . . . . . . . .3. *H. occidentalis*

      ee. Phyllaries, at least the outer, loose and spreading, with long-tapering tips; leaves not basal.

        f. Heads small, 0.5–1.0 cm wide; lower surface of leaves resin-dotted . . . . . . . . . . . . . . . . . . . . . . . . . . . . . . . . . .8. *H. microcephalus*

        ff. Heads larger, 1.5–3.0 cm wide.

          g. Leaves sessile or with very short petioles.

            h. Leaf blades clasping; stems white-villous . . . . . . . . .9. *H. mollis*

            hh. Leaf blades not clasping; stems glabrous or pubescent, not white-villous.

              i. Stems glabrous below inflorescence; leaves divaricate, subcordate to rounded at base, scabrous above, hispid to sparingly so below . . . . . . . . . . . . . . . . . . . . . . . . . . . . . .10. *H. divaricatus*

              ii. Stems pubescent below inflorescence; leaves tapering to short petioles . . . . . . . . . . . . . . . . . . . . . . . . . . . . . . .11. *H. hirsutus*

        gg. Leaves with petioles at least 5 mm long.

h. Principal leaves with 1 main nerve and several equally strong lateral nerves; stems essentially glabrous ....6. *H. grosseserratus*

hh. Principal leaves with 3 main nerves, the others not strong.

  i. Stems glabrous below the inflorescence.

    j. Leaves moderately serrate to entire, soft-pubescent below, usually short-petiolate; phyllaries only slightly exceeding the disk, never leaf-like ........................12. *H. strumosus*

    jj. Leaves deeply to moderately serrate, long-petiolate, usually attenuate at tip; phyllaries conspicuously longer than the disk, often leaf-like........................13. *H. decapetalus*

  ii. Stems hairy below the inflorescence.

    j. Upper leaves often alternate; leaves scabrous above, hirsutulous or tomentose beneath; stems scabrous or hispid to hirsute; rhizomes bearing tubers ....................14. *H. tuberosus*

    jj. Upper leaves opposite, leaves hirsute beneath, rusty or greenish, tapering to short petioles; rhizomes rarely bearing tubers .....
    ........................................11. *H. hirsutus*

1. HELIANTHUS ANNUUS L.   SUNFLOWER
Stems coarse, stout, hispid, to 3 m tall, leaves predominantly alternate, ovate, serrate, the lower cordate or subcordate, petioles long; phyllaries attenuate, most often hirsute-hispid. Widely distributed in disturbed sites. July–Sept.

The species is composed of both weedy and cultivated forms which may hybridize with several other sunflower species. Heiser et al. (1969) propose an informal working classification for the species (rather than the formal infraspecific taxa of Heiser's earlier, 1954, work) that appears to serve most purposes. In this classification, *Helianthus annuus* var. *macrocarpus* (DC.) Cockerell is the name assigned to the giant, single-headed sunflower with orange-yellow ray flowers that is often cultivated for seeds or fodder. Other forms of this variety are cultivated as ornamentals and, according to Heiser (1954), vary in number and/or color of ray flowers, the colors ranging from primrose to chestnut or plum.

All other elements of *Helianthus annuus*, the "uncultivated" plants, are separated into six groups based on various morphological features, but assigned no formal taxonomic rank. For a further discussion of the infraspecific taxa and a complete synonymy, see Heiser et al. (1969).

2. HELIANTHUS PETIOLARIS Nutt.   PRAIRIE SUNFLOWER
Stems short for sunflowers of this area, rarely over 1 m tall, pubescent with appressed hairs; lower leaves usually deltoid; ray flowers to 30, anthers generally purple. A species of more western regions of the United States, found in Ohio as a weed in dry, waste places. June–Sept.

Ohio plants belong to ssp. *petiolaris*. One specimen in The Ohio State

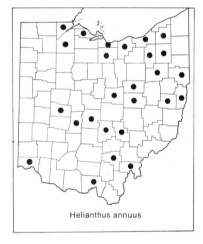

Helianthus annuus

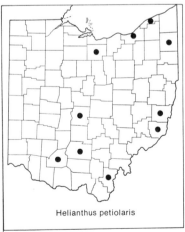

Helianthus petiolaris

x ½
Helianthus annuus

x ½

x ½
Helianthus occidentalis

Helianthus petiolaris

Helianthus angustifolius

× ½

Helianthus maximilianii

× ½

× ½

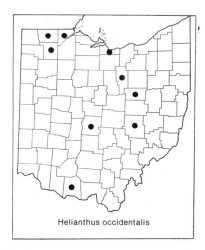

Helianthus occidentalis

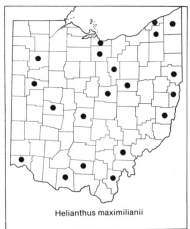

Helianthus maximilianii

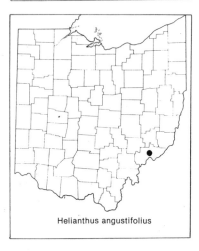

Helianthus angustifolius

University Herbarium appears to be a hybrid between this species and *H. annuus.*

### 3. **Helianthus occidentalis** Riddell  Few-leaved Sunflower

Stolons frequently present; stems scapose or with a pair or two of small, reduced leaves; leaves opposite, mostly basal or nearly so, oblong-lanceolate, entire or barely serrate, mostly scabrous above, petioles long; phyllaries imbricate; ray flowers a deep, rich yellow. Occurring primarily in dry, usually sandy, soils. July–Oct.

Although originally described from Tuscarawas County, Ohio, in 1836, by John Riddell, this beautiful sunflower is not well represented in the Ohio flora. The species reaches its greatest distribution in the central and northern Great Plains region. It is known to hybridize naturally with *H. mollis.* Ohio plants are best referred to ssp. *occidentalis.*

### 4. **Helianthus maximilianii** Schrader

Stems stout, scabrous; leaves mostly alternate, conduplicate, to 20 cm or more long, entire, sessile or nearly so; phyllaries long-attenuate, white-canescent. Ohio lies well within the range of *H. maximilianii,* but the species is represented here by only scattered collections. Dry fields and open places. July–Sept.

### 5. Helianthus angustifolius L.

This species is well marked by its narrow, revolute-margined, sessile, linear to linear-lanceolate leaves, which are rarely as much as 5.0 mm wide. It is known from only one location in Ohio, an old field in Washington County, where it was perhaps introduced (Long, 1958). July–Oct.

### 6. **Helianthus grosseserratus** Martens  Sawtooth Sunflower

Stems usually glabrous and glaucous below inflorescence or at most sparingly pubescent with appressed hairs; leaves alternate, petiolate, lanceolate to ovate-lanceolate, coarsely serrate to nearly entire, usually densely puberulent to tomentose beneath; phyllaries loose, spreading, ciliate on margins, usually tomentose on the back. Well-represented in the Ohio flora, especially along roadways and in wet prairie habitats. July–Oct.

Among all the sunflowers in Ohio, *H. grosseserratus* and *H. giganteus* are the most troublesome taxonomically. Long (1961) has shown that hybrids between these species are common, so much so that it is remarkable that the two species have not been merged. Introgression has apparently occurred, producing weedy races in many parts of the range, including Ohio. This often results in difficulty in identification of individual specimens. Even so, these two species can usually be satisfactorily separated by a combination of characters, especially the glabrous stems and the lanceolate, usually serrate, leaves with long petioles and tomentose under surfaces, which characterize *H. grosseserratus.*

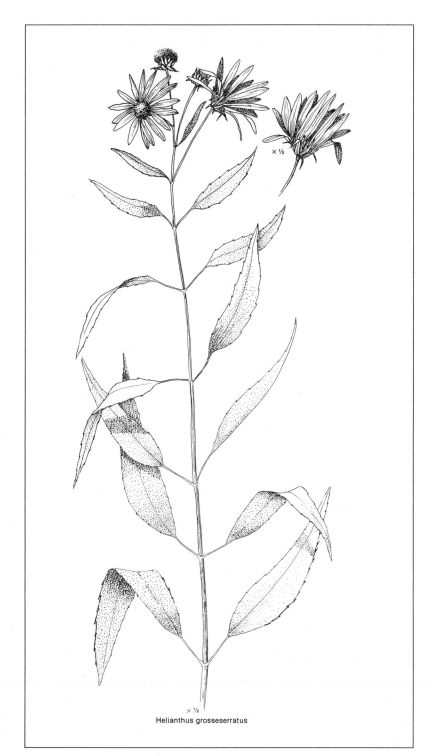

× ½

× ½
Helianthus grosseserratus

Helianthus grosseserratus

Tribe V. Heliantheae 147

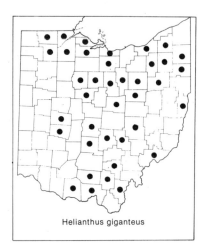

Helianthus giganteus

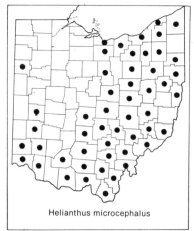

Helianthus microcephalus

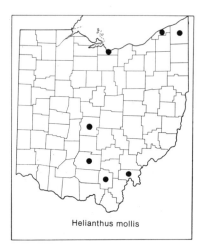

Helianthus mollis

### 7. **Helianthus giganteus** L.

Stems often reddish, pubescent below the inflorescence; leaves nearly sessile, alternate, lanceolate to ovate-lanceolate, shallowly dentate to entire, scabrous above, somewhat hirsute beneath, petioles often ciliate; phyllaries spreading, ciliate, the surface mostly glabrous or short-pubescent. These characters, used in combination, will usually permit ready identification of this species. It grows in both dry and wet habitats. Aug.–Oct.

### 8. **Helianthus microcephalus** T. & G.   SMALL WOOD SUNFLOWER

Stems glabrous, often glaucous; leaves opposite along most of the stem, alternate above, lanceolate to ovate, rounded to mostly cuneate at base, usually long-petiolate, margins nearly entire or slightly serrate, leaves scabrous above, 3-nerved; heads usually many, small. Very common in open, partially wooded habitats and similar shaded places. July–Sept.

This species hybridizes with *H. divaricatus* at many sites (Smith and Guard, 1958; Fisher, 1981), often producing hybrid swarms. The hybrid has been named *H. x glaucus* Small.

### 9. **Helianthus mollis** Lam.   HAIRY SUNFLOWER. ASHY SUNFLOWER

Stems hirsute to villous; leaves predominantly opposite, ovate to ovate-lanceolate, rounded to cordate at base, sessile, often clasping, densely cinereous-hispid or tomentose, presenting an ashy-gray appearance; phyllaries densely hispid to villous. An attractive sunflower of dry, prairie-like habitats. Not well represented in the Ohio flora, as the center of distribution is more western (in Indiana, Illinois, Missouri, Arkansas, and Tennessee). July–Sept.

### 10. **Helianthus divaricatus** L.   WOODLAND SUNFLOWER

Stems glabrous to hispid-hirsute, often monocephalic; leaves opposite, divaricate, most often triangular-ovate, rounded to subcordate to truncate at base, nearly sessile, scabrous above, sparingly hispid beneath, mostly entire. In dry, open habitats, widely distributed in Ohio. July–Oct.

### 11. **Helianthus hirsutus** Raf.

Stems stout, scabrous to densely hirsute; leaves mostly opposite, ovate, rounded or tapering to short petiole, subentire to serrate. Rather common in Ohio, where it occurs in a variety of habitats, mostly in dry, open areas. July–Oct.

Our specimens should probably be referred to var. *hirsutus*. Heiser et al. (1969) suggest that this species is closely related to *H. strumosus* and possibly should be combined with it in a single species. However, their present treatment assigns those plants with hirsute or densely scabrous stems to *H. hirsutus*, and those with glabrous stems to *H. strumosus*. This distinction works fairly well for Ohio plants. July–Oct.

x ½
Helianthus giganteus

x 1

x ½
Helianthus microcephalus

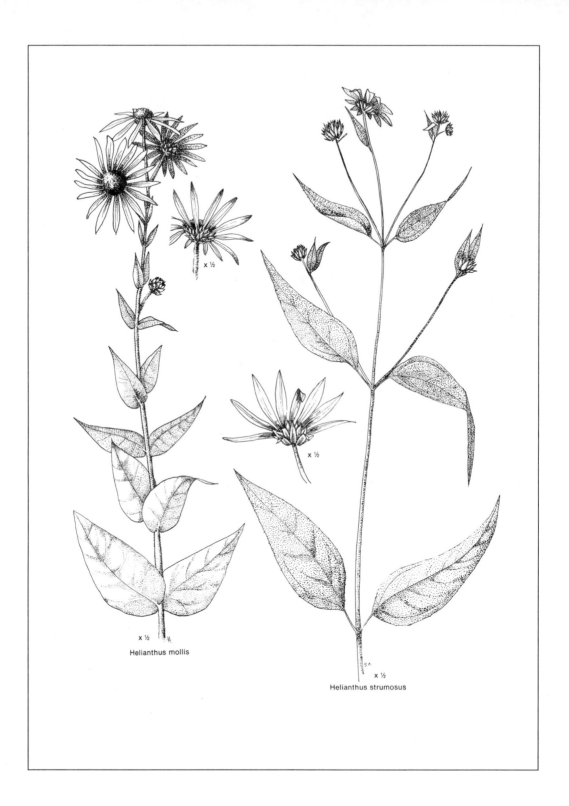

x ½

Helianthus mollis

x ½

x ½

x ½

Helianthus strumosus

x ½

x ½

x ½

Helianthus hirsutus

S. A.
x ½
Helianthus divaricatus

## 12. **Helianthus strumosus** L.   PALELEAF SUNFLOWER

Stems stout, glabrous, often glaucous, usually branched; leaves predominately opposite, linear-lanceolate to ovate, subcordate to cuneate at base, margins entire to serrate, the leaves scabrous above, glabrous to tomentulose beneath, petioles to 3 cm long; phyllaries glabrous to hispidulous on back, ciliate on margins. Rather common in wooded habitats or open places along roadways. July–Oct.

Although Ohio lies near the center of its distribution, it is difficult to establish clear characters for this variable species, the variation apparently due to hybridization with several other species.

## 13. **Helianthus decapetalus** L.

Stem branched, glabrous below the inflorescence; leaves opposite below, alternate above, ovate to ovate-lanceolate, mostly tapering at base to a somewhat winged petiole, serrate, often deeply so, scabrous above, mostly glabrous beneath; phyllaries long, loose, exceeding the disk, often leafy. Common in Ohio, growing in semishaded areas along creeks, ditches, and roadways, rarely in open places. July–Oct.

This species is composed of both diploids and tetraploids, but there has been no satisfactory method devised for separating the two morphologically. Also, this species is difficult to separate from both *H. hirsutus* and *H. strumosus*, but the more serrate leaves, long petioles and phyllaries, and pale yellow ray flowers usually are reliable distinguishing features. According to Heiser and Smith (1960), this species has hybridized with *H. annuus* to produce the cultivated taxon, *H.* x *multiflorus* L.

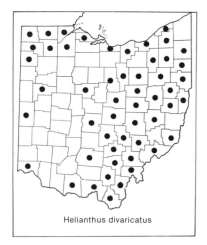

Helianthus divaricatus

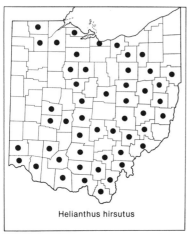

Helianthus hirsutus

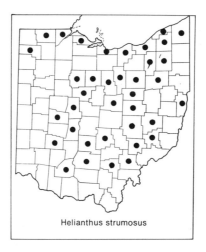

Helianthus strumosus

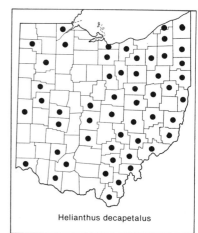

Helianthus decapetalus

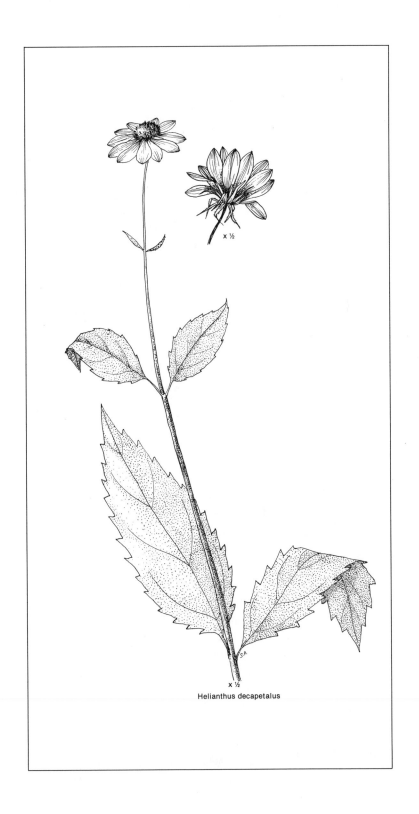

x ½

x ½

Helianthus decapetalus

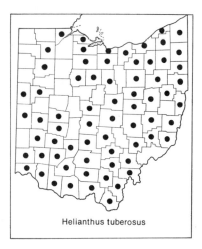

Helianthus tuberosus

## 14. **Helianthus tuberosus** L.  Jerusalem Artichoke

Stems coarse, tall, scabrous, hispid to hirsute, branched; the lower leaves opposite, the upper alternate, ovate to oblong-lanceolate, scabrous above, tomentose beneath; phyllaries ciliate on margins, hispidulous on back; rhizomes bearing edible tubers, for which this species is often cultivated. Widely distributed in Ohio, in open moist habitats. Aug.–Oct.

It is frequently difficult to separate this species from *H. strumosus*, with which it is known to intergrade (Heiser et al., 1969). In *H. tuberosus*, the stems tend to be more pubescent, more of the leaves are alternate, the blades are more serrate, broader, more decurrent on the petiole and often more pubescent beneath, and the phyllaries are darker and the rays usually larger. Natural hybridization of this species with *H. rigidus* (Cass.) Desf. produces *H. x laetiflorus* Pers. (see Heiser, 1960), which is represented in Ohio by four collections from widely separated counties. *Helianthus rigidus* itself has not been found in Ohio, but is known from nearby sites in Indiana and Michigan (Heiser et al., 1969).

### Interspecific Hybrids

Natural hybridization in *Helianthus* has been demonstrated in many instances in the numerous publications of Heiser and his students, and need not be described or elaborated upon here. However, hybridization between species of sunflowers is not uncommon in Ohio, and while it may be convenient to ignore it, some method of disposition of specimens eventually needs to be made; that is, an identification or at least a tentative identification must be determined for each specimen.

There are two methods of referring to hybrids. First, a hybrid taxon can be given a name, e.g., *Helianthus x laetiflorus*, thus conforming to the binomial system of nomenclature. The second method is the designation of a hybrid taxon by a formula, merely citing the hybridizing species, e.g., *Helianthus giganteus x H. mollis* (see Rowley, 1961, 1964).

The hybrid between *Helianthus grosseserratus* and *H. salicifolius* was originally described as a taxonomic species with the name *H. kellermanii* Britt. Later, when it was determined that the taxon was actually a hybrid, the name was altered to *H. x kellermanii*, the *x* indicating the hybrid status of the taxon. The original specimen of *H. kellermanii* was collected in Columbus, Ohio (Britton, 1901). For further discussion of this hybrid, see Kellerman (1902) and Long (1955).

As Heiser et al. (1969) point out in their excellent monograph of the genus *Helianthus*, hybridization is often an annoyance to the taxonomist for it poses problems as to what nomenclatural designation should be used, especially for

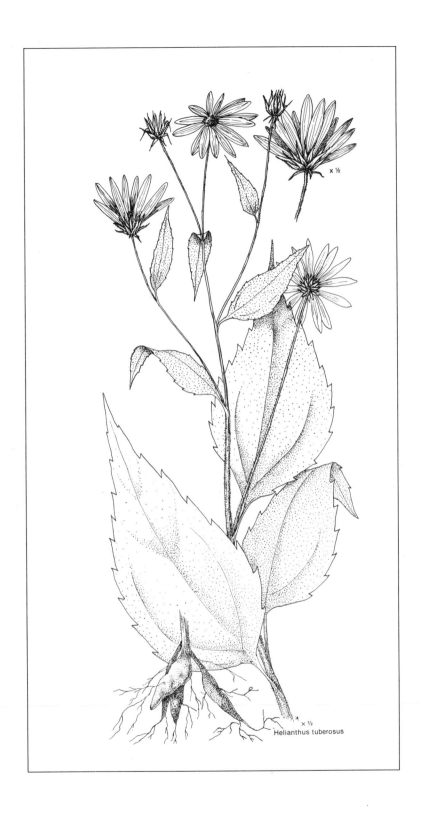

x ½

x ½
Helianthus tuberosus

filing purposes. The *International Code of Botanical Nomenclature* (Voss, 1983) permits the treatment of hybrids in either of the two manners described above. For this reason, there are binomials that have been given to many naturally occurring hybrids. While I prefer to use formulae for the designation of such hybrids, many binomials exist in the literature, and therefore these names are listed in this treatment. Hybrids are not included in keys to species, primarily because of their extreme variability.

Named hybrids of *Helianthus* that occur or may occur in Ohio:

**H.** x **ambiguus** (T. & G.) Britt. (designated by Long, 1954). *H. divaricatus* x *H. giganteus*

**H.** x **brevifolius** E. E. Wats. (designated by Jackson and Guard, 1957). *H. grosseserratus* x *H. mollis*

**H.** x **cinereus** T. & G. (designated by Jackson and Guard, 1957). *H. mollis* x *H. occidentalis*

**H.** x **divariserratus** Long (designated by Long, 1954). *H. divaricatus* x *H. grosseserratus*

**H.** x **doronicoides** Lam. (designated by Jackson, 1956). *H. giganteus* x *H. mollis*

**H.** x **glaucus** Small (designated by Smith and Guard, 1958). *H. divaricatus* x *H. microcephalus*

**H.** x **intermedius** Long (designated by Long, 1954). *H. grosseserratus* x *H. maximilianii*

**H.** x **kellermanii** Britt. (designated by Long, 1954). *H. grosseserratus* x *H. salicifolius*

**H.** x **laetiflorus** Pers. (designated by Clevenger and Heiser, 1963). *H. rigidus* x *H. tuberosus*

**H.** x **luxurians** E. E. Wats. (designated by Long, 1954). *H. giganteus* x *H. grosseserratus*
Originally described from Cedar Point, Ohio.

**H.** x **multiflorus** L. (designated by Heiser and Smith, 1960). *H. annuus* x *H. decapetalus*

With so much known about the species in this genus, it may appear that hybridization is rampant and that many species boundaries are unclear. On the contrary, the Ohio species preserve their genotypes and phenotypes quite well in nearly all instances. For example, *H. microcephalus* and *H. divaricatus* hybridize frequently where populations of the two species are contiguous or overlap. At one location in southern Ohio, these two species form a large population very typical of a hybrid swarm (Fisher, 1981). However, the surrounding populations of these two species have remained relatively "pure" at least within 5 miles of the hybrid swarm.

## 37. Verbesina L. WING-STEM. CROWN-BEARD

Heads radiate or rarely the rays absent and the heads discoid. Phyllaries few, often subequal, herbaceous, imbricated. Receptacle convex to conic, chaffy, pales stiff, embracing achenes. Disk flowers perfect, seed-forming. Ray flowers white or yellow, carpellate and seed-forming or neutral, sometimes few or absent. Pappus variable, of 1–3 rigid or weak, persistent or deciduous awns. Achenes often flattened, often winged. Leaves opposite or alternate.

For hybridization studies, see Coleman (1971, 1974, and 1977).

a. Leaves opposite; ray flowers yellow ................1. *V. occidentalis*
aa. Leaves alternate.
  b. Ray flowers 1–5, white; heads many, small ............2. *V. virginica*
  bb. Ray flowers yellow, usually more than 5.
    c. Heads 1–few, broad; rays 8–15; phyllaries spreading, erect; leaves sessile or broadly petiolate.........................3. *V. helianthoides*
    cc. Heads many, globose when mature; rays (2–) 5–10; phyllaries deflexed; leaves tapering to a short petiole ................4. *V. alternifolia*

### 1. Verbesina occidentalis (L.) Walt. YELLOW CROWN-BEARD

Stems coarse, to 2 m tall, glabrous below, puberulent above and in inflorescence; leaves opposite, ovate to ovate-lanceolate, serrate, strigose to scabrous above and beneath, petioles more or less winged, decurrent on the stem; heads small, 3–5 mm wide, numerous (20–100), crowded in a flat-topped, corymb-like arrangement; phyllaries appressed, imbricate, puberulent; ray flowers carpellate, seed-forming; achenes flattened, wingless. An uncommon species of thickets and woodland borders. Aug.–Oct.

### 2. Verbesina virginica L. TICKWEED. FROSTWEED

Stems leafy, to 2 m tall, puberulent, the hairs spreading or appressed; leaves alternate, ovate to ovate-lanceolate, serrate to nearly entire, scabrous above, usually downy-pubescent beneath, petioles more or less winged, often decurrent on the stem; heads small, rarely 1 cm wide, many (10–100); phyllaries appressed, imbricate, puberulent; ray flowers white, carpellate, seed-forming; achenes flattened, winged or wingless. Rare, recorded in Ohio only from Licking County. Open places. Aug.–Oct.

### 3. Verbesina helianthoides Michx.

Stems less than 1 m tall, hirsute, winged; leaves coarsely hirsute above, finely hirsute beneath, often densely so above and beneath, alternate, ovate to ovate-lanceolate, acute, serrate, sessile or broadly petiolate, decurrent; heads few (1–10), to 1.5 cm wide; inflorescence rather compact; phyllaries erect, loose; ray flowers yellow, carpellate, seed-forming or not; achenes winged or wingless, spreading but not reflexed in fruit. An uncommon species of dry, open places and open woodlands, mainly in central and southern Ohio. June–Oct.

Verbesina occidentalis

Verbesina virginica

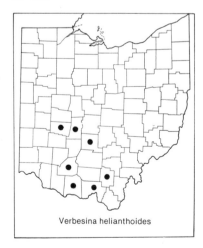

Verbesina helianthoides

Verbesina alternifolia

**4. Verbesina alternifolia** (L.) Britt.   WING-STEM

*Actinomeris alternifolia* (L.) DC.

Stems coarse, 1–2 m tall, spreading-hirsute to glabrous, usually winged by decurrent leaf bases; leaves alternate, lanceolate or elliptic-lanceolate, tapering to a short petiole, serrate to nearly entire, scabrous above, hirsute beneath, heads many (10–100 or more), in an open corymbose arrangement; phyllaries few, small, soon deflexed, especially in fruit; disk flowers spreading; ray flowers (2–) 5–10, yellow, not seed-forming; achenes spreading in fruit and forming a globose head, broadly winged or occasionally wingless; pappus of 2–3 persistent awns. Flood plains and thickets usually near streams. Common. June–Oct.

## 38. Coreopsis L.   TICKSEED

Heads radiate, long-pedunculate. Phyllaries in 2 series, the outer foliaceous and spreading, the inner broader, somewhat membranous. Receptacle flat to slightly convex, chaffy, pales thin, membranous. Disk flowers perfect, seed-forming. Pappus none, or a mere crown, or of 2 small awns. Achenes flattened, usually winged. Annual or perennial herbs with predominately opposite leaves. Often difficult to separate from *Bidens* and in our range best distinguished on the basis of characters found in the achenes and the pappus. For the basic taxonomic treatment see Sherff (1955b) and Smith (1976).

a.  Outer phyllaries much smaller and shorter than the inner; achenes wingless; disk corollas with 4 teeth........................1. *C. tinctoria*

aa. Outer phyllaries about the same length as the inner; achenes winged, although the wing often very narrow; disk corollas with 5 teeth.

b.  Leaves palmately compound or, if sessile, appearing whorled.

c.  Leaves sessile, opposite, appearing whorled ............2. *C. major*

cc. Leaves petiolate ................................3. *C. tripteris*

bb. Leaves not palmately divided, either pinnately lobed or pinnately divided.

c.  Stems leafy only below the middle, leaf segments entire or with 1 or 2 small lobes ..................................4. *C. lanceolata*

cc. Stems leafy well above the middle, the upper leaves sometimes 3-5-parted........................................5. *C. grandiflora*

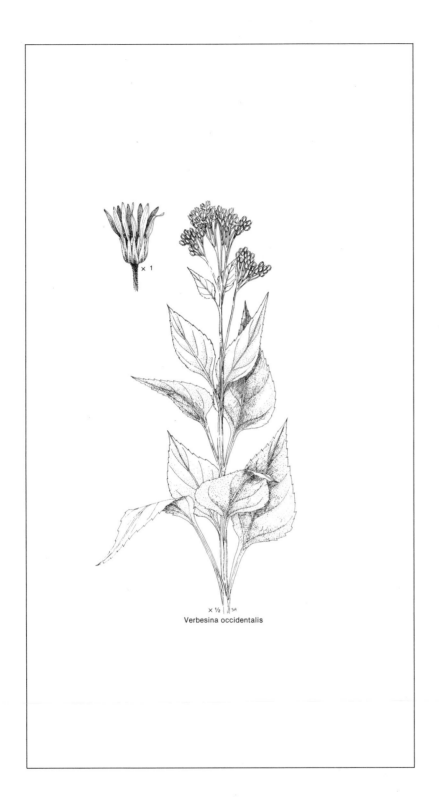

× 1

× ½

Verbesina occidentalis

Verbesina helianthoides

Verbesina alternifolia

Verbesina virginica

### 1. COREOPSIS TINCTORIA Nutt.   GOLDEN TICKSEED

Stems and leaves glabrous, annual; leaves mostly sessile, the cauline and basal pinnately divided, their segments linear or narrow-lanceolate, the upper leaves sometimes entire; heads often numerous, outer phyllaries much shorter than the inner; ray flowers yellow with a dark reddish-brown base, disk flowers deep red-purple, often becoming black in pressed specimens; achenes linear-oblong, incurved, wingless, awnless. Introduced from western states, often cultivated and frequently escaped to field-margins and roadsides. For biosystematic studies see Smith (1976), and Smith and Parker (1971). June–Aug.

### 2. **Coreopsis major** Walt.

Stems tall, minutely pubescent; leaves palmately compound, sessile, pubescent, opposite but appearing whorled; inflorescence somewhat corymbose; phyllaries hispid, the inner and the outer about the same length; disk flowers yellow; achenes elliptic-oblong. Distributed over the southern counties, where it occurs in thin woodlands and woodland borders. Often cultivated. July–Oct.

### 3. **Coreopsis tripteris** L.   TALL TICKSEED

Perennial, stems glabrous, often glaucous; leaves glabrous, usually trifoliolate, petiolate, segments ovate-lanceolate; heads in an open panicle; phyllaries mostly glabrous, the inner broader; rays yellow, disk corollas yellow, turning brown in older specimens; achenes cuneate-ovate. Widely distributed, especially in woodland borders; often cultivated. July–Sept.

C. tripteris is known to hybridize with a southeastern species, C. verticillata L., which may account for the variation I have observed in specimens collected from southeastern Ohio.

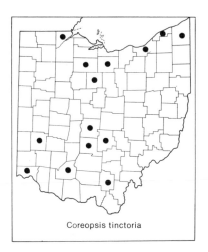

Coreopsis tinctoria

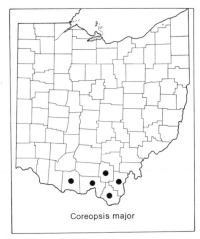

Coreopsis major

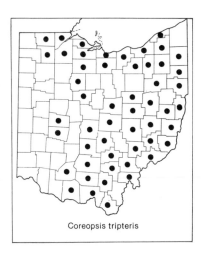

Coreopsis tripteris

× 1

× 1

× ½

Coreopsis grandiflora

× ½

Coreopsis tinctoria

× 1

× ½

Coreopsis major

× ½

Coreopsis tripteris

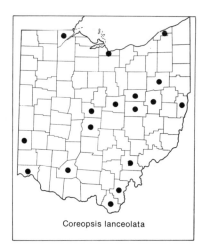

Coreopsis lanceolata

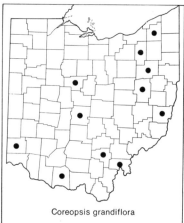

Coreopsis grandiflora

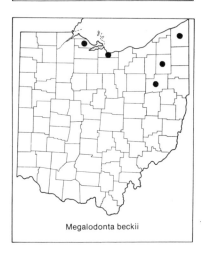

Megalodonta beckii

**4. Coreopsis lanceolata** L.

Perennial, stems tall and glabrous or pubescent at or near the base, stems leafy below; leaves simple or more frequently pinnate, glabrous or slightly villous, the lower usually long-petiolate; heads 1–few, peduncles long and naked; the outer phyllaries lanceolate, scarious, the inner ovate-lanceolate, broader than the outer, scarious; achenes with thin, wide wing, orbicular; pappus of 2 short awns. Widely distributed in dry habitats, frequently cultivated. May–July.

**5. Coreopsis grandiflora** Hogg

Apparently closely related to and hybridizing with C. *lanceolata* (Smith, 1976), but with leaves extending well up the stem; the upper leaves 3–5-parted, the segments linear-lanceolate and rarely over 5 mm wide; phyllaries lance-subulate with whitish, ciliate margins; pappus often absent. Gravelly prairies and thickets, often cultivated. May–July.

39. Megalodonta Greene    WATER-MARIGOLD

Heads radiate, on elongated stems. Phyllaries in 2 series, the outer smaller than the inner. Ray flowers yellow, neutral; disk flowers perfect, fertile. Achenes nearly terete, pappus of 3–6 long-divergent, retrorsely barbed awns. Perennial aquatic herb. Leaves opposite or whorled, the submersed leaves finely dissected, the emersed leaves lanceolate and merely serrate-margined.

Megalodonta is segregated from Bidens on the basis of evidence presented by Roberts (1985). See also study by Sherff (1955c).

**1. Megalodonta beckii** (Torr.) Greene    WATER-MARIGOLD
   *Bidens beckii* Torr.

In ponds and slow streams. In Ohio, this species is rare, possibly extirpated (Roberts and Cooperrider, 1982). July–Oct.

×½
Coreopsis lanceolata

## 40. Bidens L. BUR-MARIGOLD

Heads radiate or discoid. Phyllaries in 2 series, the outer larger and foliaceous, the inner membranous, usually striate. Receptacle chaffy, convex to nearly flat. Disk flowers perfect, seed-forming. Ray flowers yellow when present, rarely seed-forming. Pappus of 2–6 rigid awns or teeth, usually retrorsely barbed, sometimes reduced. Achenes flattened to nearly quadrangular. Annual or perennial herbs with opposite, simple to pinnately or ternately dissected leaves. For the basic treatment of the genus see Sherff (1955a).

Many species of very wide distribution, often weedy. Common in moist habitats.

a. Leaves pinnately compound or dissected.
  b. Heads discoid or with ray flowers short and rudimentary.
    c. Leaves 2–3 times pinnately divided; achenes linear, 10–18 mm long, tetragonal; heads with few flowers . . . . . . . . . . . . . . . . .1. *B. bipinnata*
    cc. Leaves trifoliolate or once pinnately compound; achenes flat.
      d. Outer phyllaries 3–5, glabrous; achenes upwardly strigose-hirsute, obscurely nerved; heads to 6 mm wide. . . . . . . . . . . . . . .2. *B. discoidea*
      dd. Outer phyllaries 5–15, sparingly to quite hairy along margins; heads larger than 6 mm wide.
        e. Outer phyllaries 5–10 (averaging 8), glabrous to sparingly pubescent on margins; achenes dark to blackish, strongly nerved, upwardly ciliate . . . . . . . . . . . . . . . . . . . . . . . . . . . . . . . . . . . . .3. *B. frondosa*
        ee. Outer phyllaries more leafy, 10–15 (averaging 13); margins usually very pubescent; achenes tan or brown, upwardly ciliate to retrorsely so at summit . . . . . . . . . . . . . . . . . . . . . . . . . . . . . . . .4. *B. vulgata*
  bb. Heads usually radiate, ray corollas longer than disk, or greatly reduced.
    c. Achenes with a thin, hispid-ciliate, brittle margin, often erose-notched, obovate.
      d. Outer phyllaries 8–12, smooth on back, margins finely ciliate; pappus awns usually divergent. . . . . . . . . . . . . . . . . . . . . . . . . . .5. *B. aristosa*
      dd. Outer phyllaries 12–14, coarsely hairy on margins, spreading or reflexed; pappus strong, curved, spreading or short . . . . . .6. *B. polylepis*
    cc. Achenes without thin, brittle margin, awns strongly subulate, divergent; outer phyllaries 6–8, ciliate or smooth . . . . . . . . .7. *B. coronata*
aa. Leaves simple, serrate, or rarely 3–5-lobed.
  b. Achene tips convex, ray flowers absent or present and large and showy; leaves sessile or connate-perfoliate; heads often nodding in fruit . . . . . .
  . . . . . . . . . . . . . . . . . . . . . . . . . . . . . . . . . . .8. *B. cernua*
  bb. Achene tips flat or slightly concave; ray flowers absent or small; leaves with petioles 1–4 cm long, the petioles often winged; heads erect in fruit . . . . . . . . . . . . . . . . . . . . . . . . . . . . . . . . . . . . . .9. *B. tripartita*

× 1

× 1

× 2

× 2

× 5

Megalodonta beckii

× ½

Bidens bipinnata

× ½

Bidens discoidea

× ½

### 1. **Bidens bipinnata** L.  SPANISH-NEEDLES

Annual, stems square, glabrous; leaves petiolate, 2 or 3 times pinnate, the terminal segments broader and rounded; heads discoid or ray-flowers small to rudimentary, disk small and apparently few-flowered; phyllaries in 2 series, the outer slightly herbaceous, green, shorter than the inner; achenes linear, 10–18 mm long, not obviously 4-sided, usually dark, nearly black at maturity; pappus of usually 4, yellow, retrorsely barbed awns. Rather common along rocky or gravelly roadsides in woods and in waste places. Aug.–Oct.

### 2. **Bidens discoidea** (T. & G.) Britt.

*Coreopsis discoidea* T. & G.

Stems slender, commonly red above, glabrous, amply branched; leaves petiolate, trifoliolate, the lateral leaflets nearly sessile, the terminal ones petiolulate, coarsely serrate; heads abundant, small, rarely 6 mm wide, 15–20-flowered; phyllaries linear to spatulate, leaf-like, exceeding the disk, essentially glabrous, at least on back; achenes dark, almost black, upwardly strigose, truncate at summit; pappus of 2 upwardly hairy awns, or the pappus sometimes much reduced. For a study of allozyme variation see Roberts (1983). Wet sandy places; not common. Aug.–Sept.

### 3. **Bidens frondosa** L.  BEGGAR-TICKS. STICK-TIGHT

Annual; stems glabrous or essentially so; leaves trifoliolate or pinnately compound, leaflets 3–5, serrate, glabrous to minutely pubescent, the lateral short-petiolulate, the terminal leaflet long-petiolulate; heads terminal, solitary; outer phyllaries 5–8, loosely ascending, herbaceous, glabrous but more often sparingly ciliate on margins, the inner 6–12, shorter and blunter; achenes upwardly soft-pubescent, median nerve strong on each face, awns erect to slightly divergent, retrorsely barbed. Common in wet, waste habitats. June–Oct.

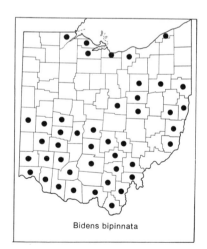

Bidens bipinnata

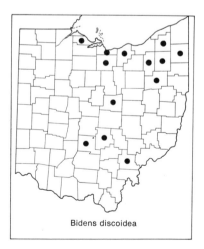

Bidens discoidea

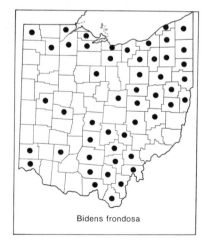

Bidens frondosa

× 2

× 1

× 2

× ½
Bidens frondosa

× ½
Bidens vulgata

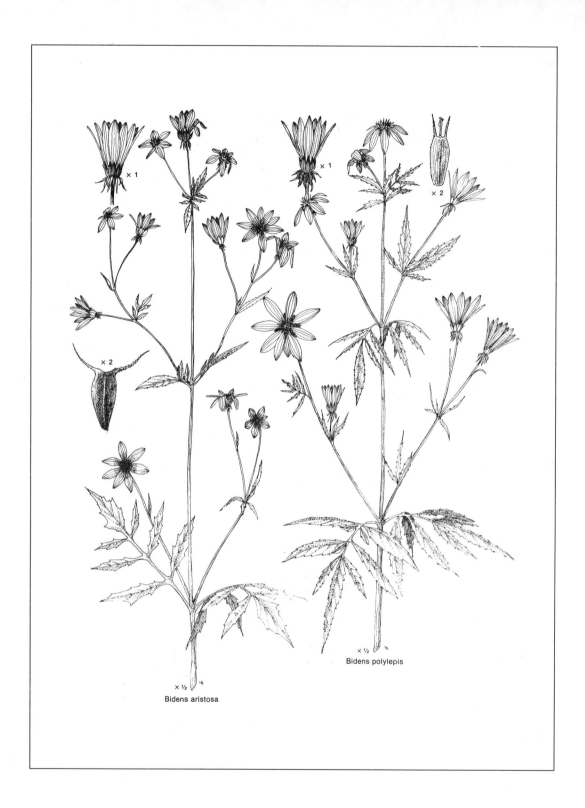

× 1

× 2

× 2

× ½

Bidens aristosa

× 1

× 2

× ½

Bidens polylepis

**4. Bidens vulgata** Greene  BEGGAR-TICKS

Coarse annual; stems to 1.5 m tall, robust, usually reddish, often furrowed and with spreading or ascending branches; leaves 3–5-foliolate, larger than those of *B. frondosa,* terminal leaflet petiolulate, the lateral ones less so; heads large, outer phyllaries 10–15, loose, leafy, subequal, ascending, copiously pubescent on margins, the inner shorter than the outer; disk olive to brown; achenes flat, midrib slender, glabrous or with a few hairs, obovate, awns erect, retrorsely barbed. Common in moist habitats and in waste places and cultivated fields, where it is often a troublesome weed. Late summer–fall.

**5. Bidens aristosa** (Michx.) Britt.  TICKSEED-SUNFLOWER

Annual; stems glabrous or slightly pubescent; leaves pinnate or seldom bipinnate, petiolate, slightly pubescent beneath; heads radiate; outer phyllaries shorter or at most the same length as the inner, glabrous to subglabrous, margins slightly ciliate to glabrous; achenes flat, sparingly pubescent, margins commonly with a narrow, brittle, erose wing, awns often divergent, long, retrorsely barbed. Scattered distribution, occurring mostly in moist habitats. Late summer–fall.

**6. Bidens polylepis** Blake

Generally similar to *B. aristosa* but more robust and showy. Outer phyllaries 12–14, slightly longer than the inner, coarsely hairy on margins and often hairy on the back; pappus of well-developed retrorsely barbed awns or awns occasionally mostly reduced. A species with scattered distribution in moist, open places. Late summer–fall.

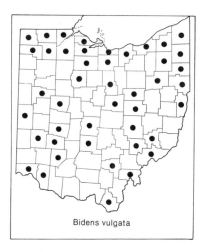

Bidens vulgata

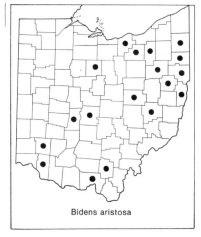

Bidens aristosa

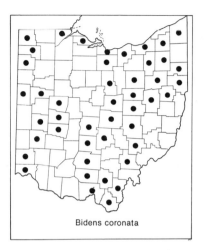

Bidens coronata

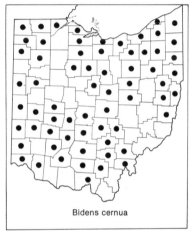

Bidens cernua

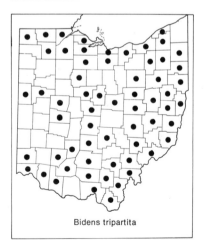

Bidens tripartita

**7. Bidens coronata** (L.) Britt. TICKSEED-SUNFLOWER
Stems slender, to 1.5 m tall, annual, glabrous; leaves short-petiolate, to 2 cm, 3–5-pinnate, segments incised or coarsely serrate; heads radiate, few to many, showy; phyllaries 6–8, about the same length as head, somewhat spatulate, smooth or with slightly ciliate margin; achenes smooth or minutely hairy, without brittle margin, the outer oblong to obovate; pappus of 2 sharp, subulate awns, minutely upwardly barbed. Widely distributed in wet prairies, swales, and flood plains. Late summer–fall.

**8. Bidens cernua** L. NODDING STICK-TIGHT
  incl. *B. laevis* (L.) B.S.P.
Annual; stems freely branching, essentially glabrous or with a few hispid hairs; leaves simple, glabrous, sessile, often connate, usually coarsely serrate; heads radiate or occasionally discoid, large, often to 5.0 cm wide, hemispheric, often nodding; outer phyllaries leafy, spreading, most often exceeding the disk, commonly reflexed; achenes compressed but quadrangular, awns usually 4, erect, retrorsely barbed. Widely distributed in low, wet habitats. Aug.–Oct.

I have found *B. laevis* inseparable from *B. cernua*, at least in our range, and therefore it has been included within this species. Possibly, intergradation has been so widespread that species lines have been nearly obliterated. As is usual in such situations, many varieties of dubious merit have been described.

**9. Bidens tripartita** L. BEGGAR-TICKS (of Europe)
  incl. *B. comosa* (Gray) Wieg. and *B. connata* Muhl.
Annual; stems glabrous; leaves simple, coarsely serrate, or occasionally lobed, petioles 1–4 cm long and narrowly winged; heads hemispheric, erect, discoid or occasionally with rays, 10–20 mm wide; outer phyllaries leafy and much exceeding the head; achenes slightly obovate, compressed but quadrangular or nearly flat, glabrous to sparingly pubescent, dark brown, to 8 mm long, awns 3–4, retrorsely barbed. A common weed of moist, waste places. Aug.–Oct.

*Bidens comosa* (Gray) Wieg. and *B. connata* Muhl. are consistently inseparable from the Old World *B. tripartita* L., and in this treatment are included within *B. tripartita*. Specimens referred to *B. connata* tend to have quadrangular achenes, while the achenes of *B. comosa* are usually more flattened.

## 41. Galinsoga Ruiz & Pavón

Heads radiate, small. Phyllaries in 2 series, membranous to herbaceous. Receptacle conic, chaffy, pales flat and narrow. Disk flowers perfect, yellow. Ray flowers 4 or 5, slightly longer than the disk, white or rarely pink, carpellate and seed-forming. Pappus of numerous laciniate scales, often fimbriate, or entirely absent in ray flowers. Achenes quadrangular, or flattened in the ray flowers. Freely branching weedy annuals with opposite, pubescent leaves. For revision of the genus, see Canne (1977).

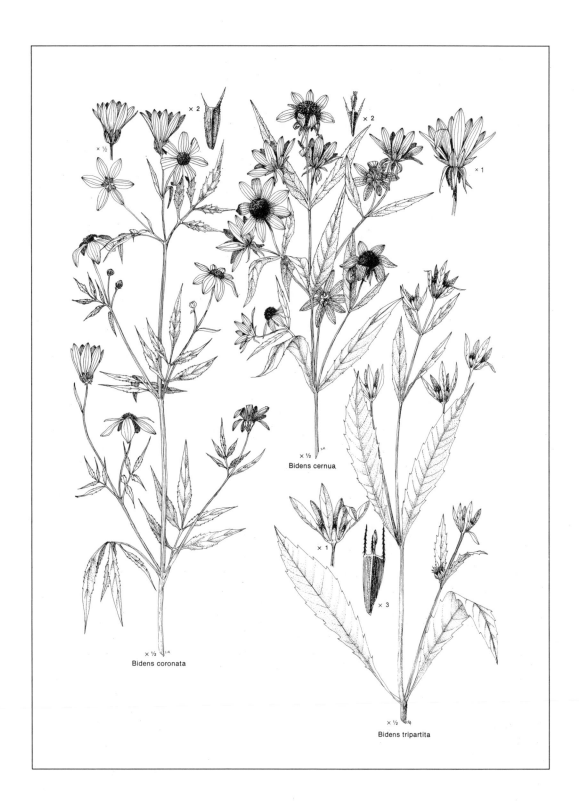

× ½

× 2

× 2

× 1

× ½
Bidens cernua

× ½
Bidens coronata

× 1

× 3

× ½
Bidens tripartita

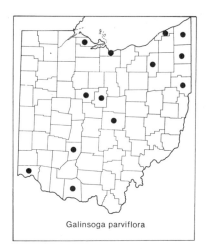

Galinsoga parviflora

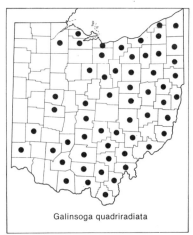

Galinsoga quadriradiata

a. Pappus of ray flowers absent, that of the disk flowers awnless . . . . . . . . .
. . . . . . . . . . . . . . . . . . . . . . . . . . . . . . . . . . . . . . . . . . . . . . .1. G. *parviflora*

aa. Pappus of ray flowers well-developed and equaling the tube, that of the
disk flowers awn-tipped . . . . . . . . . . . . . . . . . . . . . . .2. G. *quadriradiata*

1. GALINSOGA PARVIFLORA Cav.

Stems branched, glabrous to slightly pubescent with mostly appressed hairs; leaves narrowly ovate, petiolate; peduncles usually finely villous; heads numerous, aggregated in leafy cymes, small, barely 5 mm wide; ray flowers white, essentially without pappus; disk flowers with fimbriate pappus, about as long as the corolla tube, not awn-tipped; achenes sparsely pubescent or glabrous. Weed of gardens, fields, and waste places; native to tropical America. Not as common as G. *quadriradiata*. June–Nov.

2. GALINSOGA QUADRIRADIATA Ruiz & Pavón

   G. *bicolorata* St. John & White

   G. *caracasana* (DC.) Schultz-Bip.

   G. *ciliata* (Raf.) Blake

Stems nearly glabrous but more commonly spreading-hispid, often glandular; leaves ovate, coarsely toothed (more so than in G. *parviflora*); heads small, numerous, aggregated in leafy cymes, peduncles spreading-hirsute, the hairs sometimes gland-tipped; phyllaries 4–5, ovate to linear and thin; ray flowers 3–5, white, occasionally pink, 3-toothed; achenes dark, with spreading to appressed pubescence; pappus of disk flowers fimbriate, awn-tipped; pappus of ray flowers fimbriate and well-developed. Very common weedy species of gardens, fields, and waste places; native to Central America. June–Nov.

TRIBE VI. HELENIEAE CASS.

a. Leaves opposite, pinnately divided . . . . . . . . . . . . . . . . . . . .44. *Dyssodia*

aa. Leaves alternate or basal.

  b. Leaves basal . . . . . . . . . . . . . . . . . . . . . . . . . . . . . . . . . .42. *Hymenoxys*

  bb. Leaves cauline, alternate . . . . . . . . . . . . . . . . . . . . . . . . . .43. *Helenium*

42. Hymenoxys Cass.

Heads radiate, large, showy, solitary on leafless scapes. Phyllaries in 2–3 series, somewhat imbricate, their margins scarious. Receptacle slightly convex, naked. Disk flowers yellow, perfect, seed-forming. Ray flowers carpellate, seed-forming, yellow. Pappus of scales, often aristate. Achenes somewhat turbinate, 5-angled, pubescent. Low perennial herbs with basal, entire, linear to linear-oblanceolate leaves.

About 20 species of scapose or leafy-stemmed herbs native chiefly to western North America and South America.

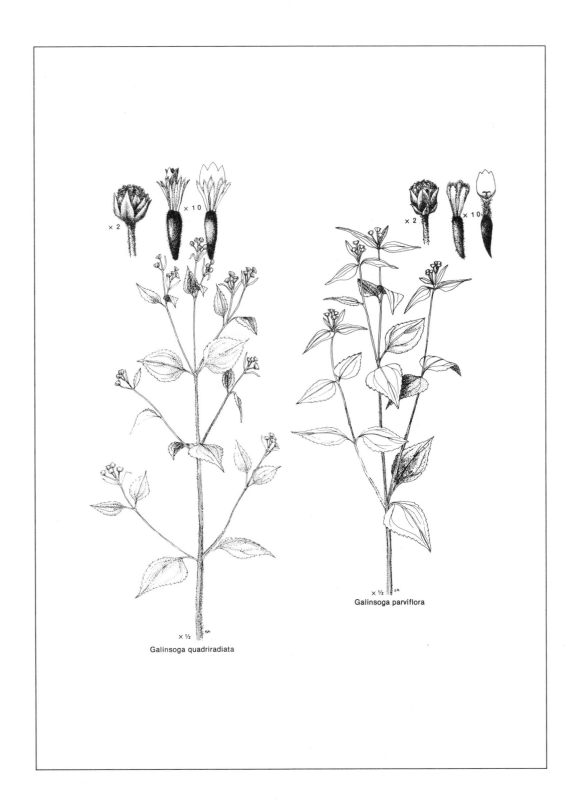

×2

×10

×2

×10

× ½

Galinsoga parviflora

× ½

Galinsoga quadriradiata

Hymenoxys acaulis

1. **Hymenoxys acaulis** (Pursh) K. Parker   Lakeside Daisy

*Actinea acaulis* (Pursh) Spreng.

*Actinea herbacea* (Greene) B.L. Robinson

*Actinella acaulis* (Pursh) Nutt.

*Tetraneuris herbacea* Greene

This species is very rare in Ohio and known only from open sites on thin, calcareous soils on Marblehead Peninsula of Ottawa County, where the first collection was made in 1890. While the population has persisted, it apparently is not spreading. Whether or not the species is native to Ohio is discussed by Weed (1890) and Moseley (1931). Our plants should be referred to as var. *glabra* (Nutt.) K. Parker. April–June.

× ½

Hymenoxys acaulis

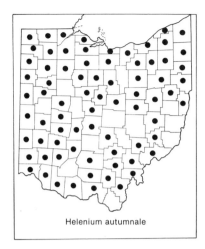

Helenium autumnale

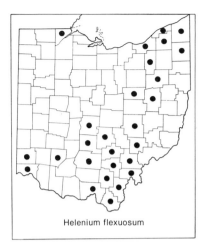

Helenium flexuosum

## 43. Helenium L.  SNEEZEWEED

Heads radiate, rarely discoid. Phyllaries in 2–3 series, subequal, reflexed. Receptacle convex to conic, naked. Disk flowers perfect, numerous, seed-forming. Ray flowers 3-lobed, carpellate and seed-forming or neutral, yellow or brownish-yellow. Pappus of thin scales, the nerve often awn-tipped. Achene 4–5-ribbed, pubescent on angles. Annual or more often perennial herbs with alternate, usually decurrent, glandular-punctate leaves.

About 50 species native to the New World; represented in Ohio by three species. See Bierner (1972) for the taxonomy of North American *Helenium*.

a.  Leaves linear to lanceolate, decurrent on stem; plants perennial.
  b.  Disk corollas yellow . . . . . . . . . . . . . . . . . . . . . . . . . . . . .1. *H. autumnale*
  bb. Disk corollas purple or brownish . . . . . . . . . . . . . . . . . .2. *H. flexuosum*
aa. Leaves linear-filiform, not decurrent on stem; plants annual . . . . . . . . . .
  . . . . . . . . . . . . . . . . . . . . . . . . . . . . . . . . . . . . . . . . . .3. *H. amarum*

### 1. **Helenium autumnale** L.  COMMON SNEEZEWEED

Perennial, stems to 1.5 m tall, glabrous to somewhat puberulent, winged; leaves abundant, narrowly ovate-lanceolate, tapering to sessile base, decurrent onto stem, toothed to subentire, glandular-punctate, the lower deciduous at or before anthesis; heads abundant on leafy branches, hemispheric; disk flowers yellow; phyllaries narrow, acuminate, puberulent, deflexed; ray flowers mostly seed-forming, yellow, somewhat drooping; pappus of awn-tipped scales. Very common along roadsides and in waste places; probably in every Ohio county. Aug.–Oct.

### 2. HELENIUM FLEXUOSUM Raf.  NAKED SNEEZEWEED

    *H. nudiflorum* Nutt.

Perennial, stems to about 1 m tall, puberulent to somewhat hirsute-

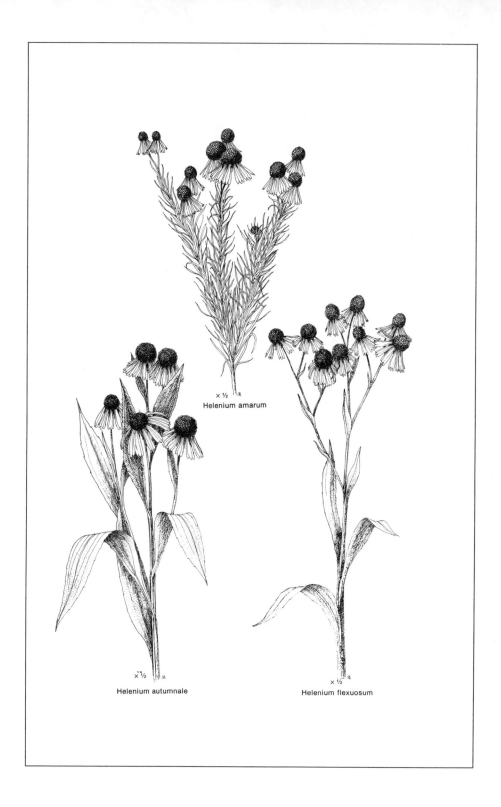

× ½

Helenium amarum

× ½

Helenium autumnale

× ½

Helenium flexuosum

puberulent, winged; leaves smaller and more linear than those of *H. autumnale*, subentire, glandular-punctate, often puberulent, sessile and decurrent onto stem, reduced upward, the lower often deciduous at anthesis; heads often fewer than in *H. autumnale*, globose, leafy-bracteate; phyllaries linear-lanceolate and mostly acuminate, puberulent, deflexed; ray flowers neutral, yellow, trilobed; disk corollas purple, brown, or brownish-purple; tip of pappus tapering to awn. For basic treatment of the species, see Rock (1957). Evidently introduced from southern states. Common in moist ground and open waste places, but not so common as *H. autumnale*. June–Oct.

3. HELENIUM AMARUM (Raf.) H. Rock   SLENDER-LEAVED SNEEZEWEED

    *H. tenuifolium* Nutt.

Annual, stems branched from near base, short, essentially glabrous; leaves numerous, linear-filiform, not decurrent onto stem, glandular-punctate, heads globose on short peduncles; phyllaries linear, punctate, spreading to deflexed; ray flowers carpellate and seed-forming, yellow, 3-lobed; disk flowers carpellate, seed-forming, yellow; pappus scales scarious, hyaline, awn-tipped. Introduced from southern and western areas of the United States. Uncommon species of disturbed habitats; reported from only four counties. June–Oct.

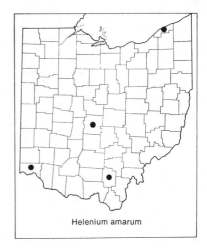

Helenium amarum

Dyssodia papposa

### 44. Dyssodia Cav.   FETID MARIGOLD

Heads radiate, rarely discoid. Phyllaries gland-dotted, in 1–2 series, the inner united at base. Receptacle essentially flat, naked or with a few weakly developed pales. Disk flowers perfect, numerous. Ray flowers few, carpellate, seed-forming, yellow or somewhat orange, barely exceeding the disk. Pappus of scales divided into bristles. Achenes striate. Ill-scented annuals with mostly linear, gland-dotted, pinnate to bipinnate, opposite leaves. For the basic taxonomic treatment of the genus, see Strother (1969).

1. **Dyssodia papposa** (Vent.) A. S. Hitchc.

An uncommon species of dry habitats. July–Oct.


## TRIBE VII. ANTHEMIDEAE CASS.

a.  Receptacles chaffy.

  b.  Receptacles nearly flat; disk corollas white; achenes oblong, flattened; heads many and arranged in a corymb.................45. *Achillea*

  bb.  Receptacles mostly conic; disk corollas yellow; achenes terete, angled or ribbed; heads solitary at tips of branches..............46. *Anthemis*

aa.  Receptacles naked (hairy in one species of *Artemisia*).

  b.  Heads discoid.

    c.  Heads small, less than 5 mm wide, arranged in spikes, racemes, or panicles......................................50. *Artemisia* cc. Heads larger, greater than 5 mm in width.

    d.  Receptacles conic; phyllaries shorter than disk . . . . . . .47. *Matricaria*

   dd. Receptacles flat or slightly convex.

      e.  Blades 1–3 times pinnately divided . . . . . . . . . . . . . .49. *Tanacetum*

     ee. Blades crenate-serrate, often lobed at base . . . . .48. *Chrysanthemum*

bb. Heads radiate.

   c.  Plants annual; receptacles conic to hemispheric . . . . . . .47. *Matricaria*

  cc. Plants perennial; receptacles flat to slightly convex . . . . . . . . . . . . . .

   . . . . . . . . . . . . . . . . . . . . . . . . . . . . . . . . . . . . . .48. *Chrysanthemum*

## 45. Achillea L. YARROW

Heads many, radiate, small; inflorescence corymbose. Phyllaries imbricate in 3–4 series, scarious or at least with hyaline margins and green midribs. Receptacle slightly convex, chaffy. Disk flowers perfect, many, seed-forming. Ray

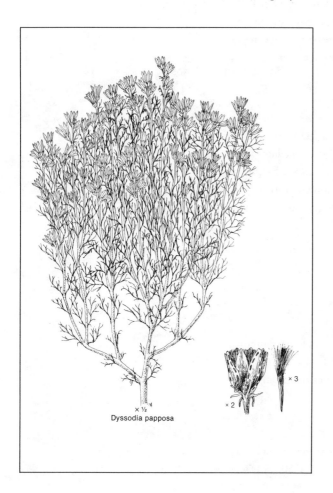

Dyssodia papposa

Achillea millefolium

flowers 5–12, carpellate, seed-forming, white to light pink or light purple. Pappus absent. Achenes flattish, glabrous. Perennial, aromatic herbs with alternate, pinnately dissected leaves. Several species are cultivated and may be found as escapes, especially A. *ptarmica* L., cited by Long (1958), which appears to be invading the midwest from the northeast.

1. ACHILLEA MILLEFOLIUM L.   YARROW

A native of Europe and Asia, well-established in Ohio as a weed in waste places and disturbed sites. June–Oct.

Achillea millefolium

## 46. Anthemis L.   CHAMOMILE

Heads radiate or occasionally discoid. Phyllaries subequal and imbricate in several series, margins hyaline. Receptacle conic to slightly convex, chaffy throughout or only toward center. Disk flowers yellow, perfect, seed-forming. Ray flowers yellow or white, carpellate, seed-forming or not. Pappus usually absent or represented by a short crown. Achenes terete or approaching 4–5-angled. Mostly aromatic, annual or marginally perennial herbs with pinnatifid or pinnately dissected, alternate leaves. Introduced and naturalized weedy species from Europe, Asia, and Africa. In addition to the species below, A. *nobilis* L., has been collected in Ohio, but is very uncommon and evidently a waif.

a.  Ligulate corollas yellow............................1. A. *tinctoria*
aa. Ligulate corollas white.
  b.  Ligulate flowers seed-forming; receptacle chaffy throughout; achenes smooth, about 10-nerved; herbage odorless ...........2. A. *arvensis*
  bb. Ligulate flowers not seed-forming; receptacle chaffy only toward center; achenes striate to rough-tuberculate; herbage strongly scented ........
    ..............................................3. A. *cotula*

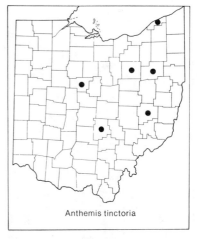

Anthemis tinctoria

1. ANTHEMIS TINCTORIA L.   YELLOW COTULA

Perennial with short stems to 40 cm tall, usually with a few branches above, puberulent to sparingly villous above; leaves pinnatifid, rachis winged, segments toothed, whitish-villous pubescent; heads solitary and usually long-pedunculate, about 15 mm wide; phyllaries villous-tomentose; rays yellow; receptacle chaffy throughout, chaff narrow with yellow awn-tips; achenes striate-nerved; pappus short, merely a crown. Naturalized from Europe, this is an attractive, often cultivated species. Open fields and waste places; not common in Ohio. June–Aug.

2. ANTHEMIS ARVENSIS L.   CORN COTULA

  A. *mixta* L.

Annual, more branched than A. *tinctoria*, odorless, somewhat puberulent or sparingly villous; leaves bipinnatifid, rachis with a very narrow wing; heads abundant, 6–12 mm wide, pedunculate, becoming globose with age; phyllaries

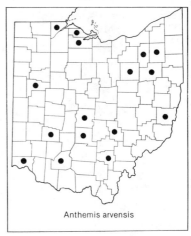

Anthemis arvensis

villous-tomentose; ray flowers carpellate and seed-forming, white; receptacle chaffy throughout, pales thin, awns cuspidate-tipped; achenes striate-nerved; pappus a short crown. Naturalized from Europe. Found along roadsides and in waste places. May–Sept.

3. ANTHEMIS COTULA L.   MAYWEED. STINKING COTULA

Annual, branched as A. *arvensis*, ill-scented, nearly glabrous, about 25–30 cm tall; leaves 2–3 times pinnatifid, segments narrow; heads numerous, short-pedunculate, about 8 mm wide, becoming somewhat cylindric with age; phyllaries less villous than those of A. *arvensis*; rays white, neutral, receptacle chaffy only in the center, pales narrow and tapering to a slight awn; achenes tuberculate, pappus none. Naturalized from Europe, a widespread weed in Ohio, where it occurs along roadsides, in fields, and in waste places. May–late fall.

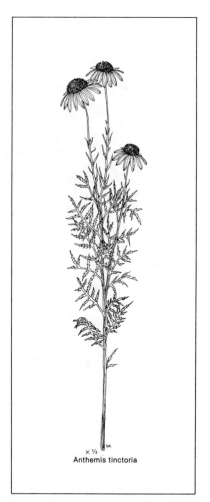

× ½

Anthemis tinctoria

× 5

× ½

Anthemis arvensis

× 1

× ½

Anthemis cotula

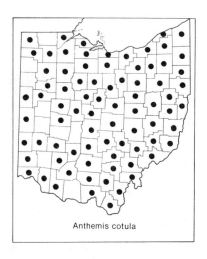

Anthemis cotula

## 47. Matricaria L. WILD CHAMOMILE

Heads radiate or discoid. Phyllaries in 2–3 series. Receptacle naked, conic to hemispheric. Disk flowers yellow, perfect, seed-forming. Ray flowers white, carpellate, usually seed-forming. Pappus absent or represented by a short crown or teeth. Achenes smooth to slightly roughened. Annual or perennial herbs with alternate, pinnatifid or pinnately compound leaves. A genus of perhaps 40 species, native to the northern hemisphere and South Africa.

a.  Heads discoid; corollas 4-lobed.................1. *M. matricarioides*
aa. Heads radiate; disk corollas 5-lobed.

b.  Receptacles hemispheric; achenes strongly 3-ribbed; plants essentially
    scentless......................................2. *M. maritima*

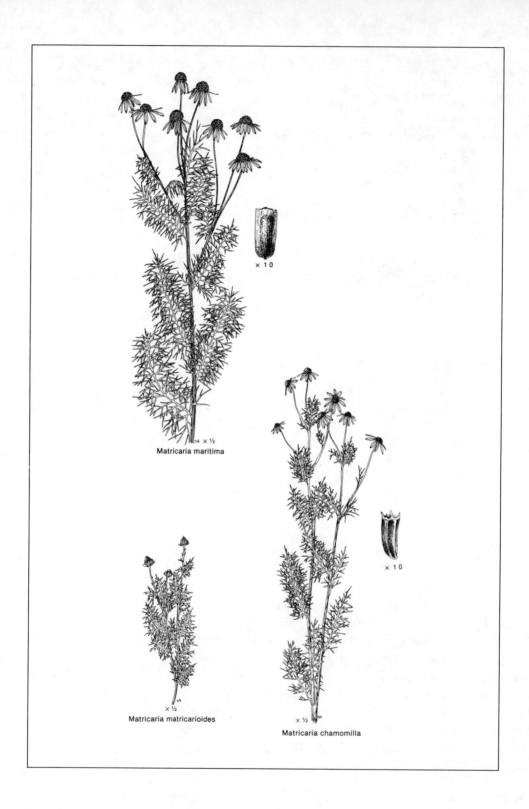

Matricaria maritima

× 10

Matricaria matricarioides

× ½

Matricaria chamomilla

× ½

bb. Receptacles conic; achenes not strongly 3-ribbed; plants scented . . . . . .
. . . . . . . . . . . . . . . . . . . . . . . . . . . . . . . . . . . . . . . . . . .3. M. *chamomilla*

**1. MATRICARIA MATRICARIOIDES (Less.) T. C. Porter   PINEAPPLE WEED**

Stems annual, to 15 cm tall, usually branching and leafy, glabrous; the heads when crushed emitting an odor of fresh pineapple, leaves 2–3 times pinnatifid; heads many, rayless; receptacle broadly conic; phyllaries with hyaline margins; disk corollas 4-lobed; achenes with 2 marginal nerves or ridges and one or more weak nerves between; pappus a short crown. Introduced from western United States, a common weedy species of rather sterile habitats. June–Oct.

**2. MATRICARIA MARITIMA L.   SCENTLESS FALSE CHAMOMILE**

   *M. inodora* L.

Scentless annual, stems to 0.5 m tall, essentially glabrous; leaves bipinnatifid, the segments linear-filiform; heads numerous; ray flowers white; disk corollas 5-lobed; receptacle hemispheric or rounded; achenes with 3 strongly thickened and somewhat wing-like ribs, the surface finely rugose; pappus a short crown. Native to Europe; found along roadsides and in waste places. Uncommon. July–Sept.

**3. MATRICARIA CHAMOMILLA L.   FALSE CHAMOMILE**

Aromatic annual; stems to 0.5 m tall, glabrous; leaves bipinnatifid, the segments linear; heads abundant; receptacle conic; rays white; disk corollas 5-lobed; achenes usually 5-ribbed, the ribs not wing-like, surface of achene smooth; pappus a short-toothed crown. Native to Europe and Asia and now introduced into the United States, but uncommon in Ohio, where it occurs in waste places and along roadsides. Late spring–fall.

**48. Chrysanthemum L.   CHRYSANTHEMUM**

Heads radiate, discoid, or appearing discoid. Phyllaries in 2–4 series, imbricate, midribs green, margins hyaline. Receptacle flat or slightly convex, naked. Disk flowers yellow, perfect, seed-forming. Ray flowers white (or yellow in some cultivated forms), often greatly reduced in one species. Pappus absent or a mere ridge. Achenes glabrous, ribbed. Perennials with alternate, entire, dentate or dissected leaves. Native to Europe and Asia, naturalized in our range. Occasionally, species other than those listed below can be found as escapes in Ohio, namely *C. coccineum* Willd., PAINTED DAISY or PYRETHRUM; and *C. segetum* L., CORN-MARIGOLD.

a.  Heads solitary, showy . . . . . . . . . . . . . . . . . . . . . . . . .1. *C. leucanthemum*
aa. Heads in clusters, smaller than above.
   b.  Leaves pinnatifid to bipinnatifid; ray flowers present, white . . . . . . . . . .
. . . . . . . . . . . . . . . . . . . . . . . . . . . . . . . . . . . . . . . .2. *C. parthenium*
   bb. Leaves crenate; ray flowers seemingly absent . . . . . . . . .3. *C. balsamita*

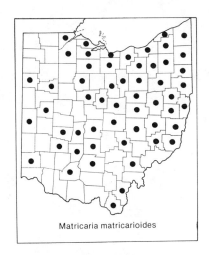

Matricaria matricarioides

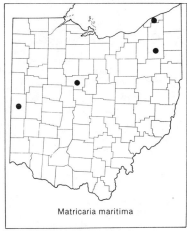

Matricaria maritima

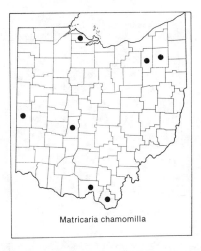

Matricaria chamomilla

### 1. CHRYSANTHEMUM LEUCANTHEMUM L.   OX-EYE DAISY. WHITE DAISY

Perennial with well-developed rhizomes; stems usually solitary, to 0.6 m tall, glabrous or with a few scattered hairs; leaves essentially glabrous, basal leaves spatulate-obovate on long petioles, crenate or more often lobed, the upper reduced and becoming sessile, oblong or oblanceolate, coarsely crenate or dentate, lower teeth often larger and spreading; heads usually solitary; phyllaries narrow, brown-margined, the outer lance-triangular; ray-flowers 15–20 (more in double forms), white; disk flowers yellow; achenes nerved, somewhat quadrangular; pappus none. An attractive and common weed naturalized from Europe and widely established in open fields, along roadsides, and in other waste places. June–Oct.

Our common Ox-eye Daisy apparently belongs to the dubious variety *pinnatifidum* Lecoq. & Lamotte.

### 2. CHRYSANTHEMUM PARTHENIUM (L.) Bernh.   FEVERFEW

Taprooted perennial; stems to 0.8 m tall, glabrous below, occasionally puberulent above; leaves puberulent beneath, glabrous above, bipinnatifid, short-petiolate or sessile; heads several, in corymbose arrangement; phyllaries pubescent on back, narrow and with hyaline tips; rays 10–20 (more in double forms), white; achenes terete to subquadrangular, ribbed; pappus at most a crown. Native to Europe. Our specimens are probably escapes from nearby gardens as this species is often used in perennial borders. I see no evidence that it is becoming widely naturalized. June–Aug.

### 3. CHRYSANTHEMUM BALSAMITA L.   COSTMARY. SWEET-MARY

Fragrant perennial, 0.5 m or more tall; stems nearly square (at least when dry), strigose-pubescent above, glabrous below; leaves silvery-pubescent above and beneath, crenate, often lobed at base, basal leaves oblanceolate and long-petiolate, the upper mostly numerous and sessile; heads many in a corymbose

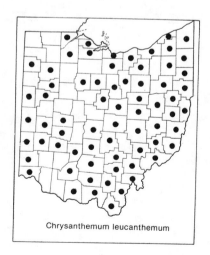

Chrysanthemum leucanthemum

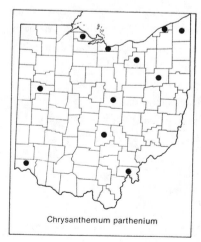

Chrysanthemum parthenium

Chrysanthemum balsamita

× ¼

Chrysanthemum balsamita

× ½

Chrysanthemum leucanthemum

× ½

Chrysanthemum parthenium

Tribe VII. Anthemideae    187

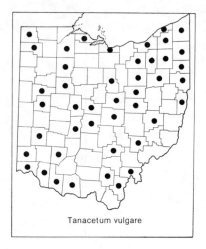

Tanacetum vulgare

arrangement; rays absent or, if present, very short; phyllaries narrow and with small hyaline tips; achenes terete to subquadrangular; pappus a mere border or slight crown. A native of Asia often grown as a culinary herb. Uncommon in Ohio; when found it is no doubt as an escape from cultivation. Aug.–Oct.

## 49. Tanacetum L.   TANSY

Heads appearing discoid, the inflorescence corymbiform. Phyllaries in several series, imbricate, margins and tips scarious. Receptacle flat or slightly convex, naked. Disk flowers perfect, yellow. Ray flowers with very short, scarcely expanded corollas, the heads thus appearing discoid. Pappus absent or represented by a short crown. Achenes somewhat glandular, 5-ribbed. Strong-scented perennial herbs with alternate, pinnately dissected leaves. Introduced from Europe and Asia and represented in Ohio by a single species.

1. TANACETUM VULGARE L.   COMMON TANSY

Introduced from Eurasia; well-established in many areas along roadsides, in ditches, and in other waste places. Midsummer–fall.

## 50. Artemisia L.   WORMWOOD. MUGWORT. SAGE

Heads small, discoid; inflorescence paniculate, spicate, or racemose. Phyllaries imbricate, scarious, receptacles flat to somewhat hemispheric, naked or pubescent. Flowers on periphery of heads carpellate, the inner flowers perfect; fruits forming throughout or only near outside. Pappus absent. Achenes obovoid to oblong. Annual or perennial, aromatic, often tomentose herbs. Leaves alternate, entire to dissected.

The genus consists of as many as 100 species. In addition to those treated below, a few waifs and adventive escapes have been collected in and reported from Ohio, but are not included here. For instance, *A. canadensis* Michx., a common Canadian species, has been found in Erie County. Others, such as *A. stelleriana* Besser, DUSTY MILLER; and *A. sacrorum* Ledeb., SUMMER FIR, have escaped from cultivation.

a.  Leaves white-tomentose or silvery-sericeous above and/or beneath.
  b.  Leaves silvery-sericeous above and beneath; receptacles hairy . . . . . . . . .
    . . . . . . . . . . . . . . . . . . . . . . . . . . . . . . . . . . . . . . . . . .1. *A. absinthium*
  bb. Leaves white-tomentose; receptacles glabrous.
    c.  Leaves 2–3 times pinnatifid; plants shrubby . . . . . . . . . . . .2. *A. pontica*
    cc. Leaves entire to once pinnatifid.
      d.  Leaves entire to remotely toothed or lobed, densely white-tomentose beneath and usually above . . . . . . . . . . . . . . . . . . . .3. *A. ludoviciana*
      dd. Leaves pinnatifid, segments entire or incised, green and glabrous above, white-tomentose beneath, the margins inrolled; often with stipule-like lobes at base . . . . . . . . . . . . . . . . . . . . . . . .4. *A. vulgaris*

aa. Leaves glabrous or pubescent, but not white-tomentose or silvery-sericeous either above or beneath.

  b.  Heads essentially sessile, crowded, the panicle narrow; seeds forming in all flowers; herbage barely scented .................... 5. *A. biennis*

 bb. Heads pedunculate; herbage sweet-scented.

   c.  Seeds not forming in central flowers; inflorescence narrow ..........
............................................. 6. *A. caudata*

  cc. Seeds forming in all flowers; inflorescence open and somewhat spreading............................................... 7. *A. annua*

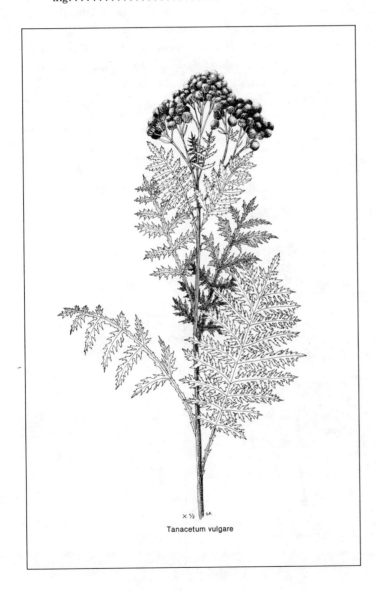

× ½    SA

Tanacetum vulgare

Artemisia absinthium

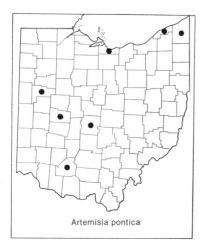
Artemisia pontica

1. ARTEMISIA ABSINTHIUM L.   WORMWOOD. ABSINTHIUM

Perennial, fragrant herb or semishrub, to 1 m tall, finely canescent; leaves silky or sericeous on both surfaces, 2–3 times pinnatifid, segments obtuse, the upper leaves reduced and less divided, on short petioles, and with segments more acute; inflorescence paniculate, leafy; heads nodding, abundant; phyllaries imbricate, scarious on margins; receptacle long-hairy among the flowers; achenes glabrous. Waste places. Midsummer–fall.

A native of Europe, this species is included here with reluctance, since it is an escape from cultivation and rarely found in Ohio; however, it is apparently becoming established, especially in Canada and in the eastern United States.

2. ARTEMISIA PONTICA L.   ROMAN WORMWOOD

Semiwoody perennial to 1 m tall; stems puberulent to glabrous; leaves numerous, crowded, grayish to white-tomentose above and beneath, 2–3 times pinnatifid, segments linear, toothed, with a pair of leafy lobes or stipules at base; heads in a somewhat narrow, elongated panicle; receptacle glabrous; phyllaries

× ½
Artemisia absinthium

Artemisia vulgaris
× ½

Artemisia biennis
× ½

Artemisia pontica
× ½

Artemisia ludoviciana
× ½

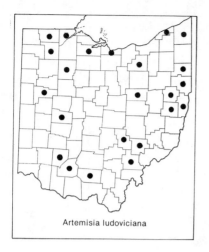

Artemisia ludoviciana

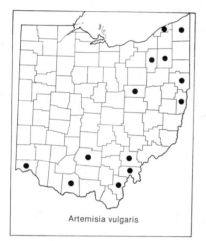

Artemisia vulgaris

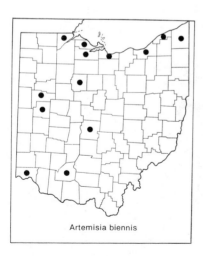

Artemisia biennis

herbaceous, densely pubescent; flowers seldom developing. Native to Europe, this species is cultivated and often escapes to roadsides and waste places. Raised as a garden herb, it is sometimes used in home medicines, especially as a vermifuge. Wormwood tea is remembered by many people who were raised in the country. Late summer–early fall.

3. ARTEMISIA LUDOVICIANA Nutt.   MUGWORT. WHITE SAGE

Aromatic perennial; stems to 1 m tall; stems white-tomentose; leaves with matted, white-felted hairs above and beneath, lance-linear to slightly lanceolate, entire or with a few coarse, irregularly spaced teeth or lobes; heads loose, globose or cylindric, in a paniculate arrangement; phyllaries tomentose on back; receptacle glabrous; achenes glabrous. Apparently, this species is rapidly becoming established in prairies and dry, waste places. Midsummer–early fall.

An extremely variable and weedy species, more common in the midwestern and western United States than it is in Ohio, this species is often segregated into several varieties. In addition to var. *ludoviciana*, with leaves usually toothed, the most common variety in our range is var. *gnaphalodes* (Nutt.) T. & G., with leaves mostly entire.

4. ARTEMISIA VULGARIS L.   COMMON MUGWORT

Aromatic perennial; stems to 2 m tall, longitudinally ridged and grooved, purplish or greenish, glabrous or mostly so below inflorescence; leaves green and glabrous above, white-tomentose beneath, petiolate, with stipule-like appendages at base, segments unequal, toothed or nearly entire, margins inrolled (at least when dry); heads in a paniculate arrangement, branches ascending, often crowded into glomerules, sessile, ovoid; receptacle glabrous; phyllaries gray-tomentulose, linear-lanceolate; achenes glabrous, not nerved. Native to Eurasia and now established in eastern United States. See study by Keck (1946). Fields and roadsides. Midsummer–early fall.

5. ARTEMISIA BIENNIS Willd.

Coarse taprooted annual with barely scented herbage; stems glabrous, to 3 m tall; leaves pinnatifid or bipinnatifid, segments acute, all but the uppermost cut-toothed; heads in a tight spiciform, leafy arrangement, numerous, short-pedunculate or sessile, subglobose; phyllaries glabrous, roundish, margins scarious, midrib green; receptacle glabrous; achenes glabrous. Waste places, especially in gravelly or sandy soil. Native to the northwestern United States but becoming well-established in the midwest and east. Aug.-Oct.

6. ARTEMISIA CAUDATA Michx.

Taprooted biennial or occasionally perennial; stems usually single, purplish to greenish, glabrous; leaves 2–3 times pinnatifid, segments linear or linear-filiform, essentially glabrous; heads in a rather elongate and narrowly paniculate arrangement; heads crowded, short-pedunculate, nodding with age;

phyllaries glabrous, margins scarious, otherwise green. Rather common in Erie County but scarce elsewhere. Occurs in dry, sandy habitats. July–Sept.

7. Artemisia annua L.    Sweet Wormwood

Annual, sweet-scented, glabrous; stems to nearly 1.5 m tall, often branched; leaves 2–3 times pinnatifid, the segments narrowly oblong to lanceolate, sharply toothed; heads nodding, pedunculate, in an open, somewhat spreading, paniculate arrangement; phyllaries glabrous, scarious except for green midrib; achenes narrow, glabrous, turbinate. A weedy species of open fields and waste places, native to Europe and Asia but well-established in Ohio. Aug.–Oct.

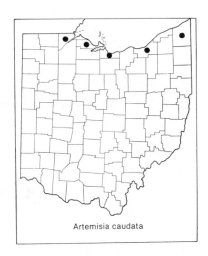

Artemisia caudata

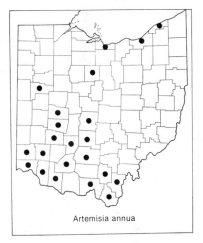

Artemisia annua

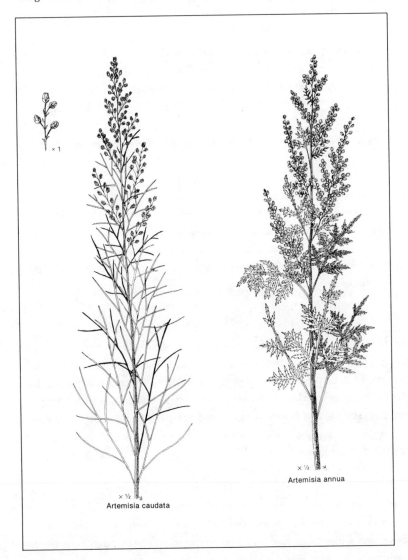

Artemisia annua

Artemisia caudata

# TRIBE VIII. SENECIONEAE CASS.

a. Heads discoid.

  b. Phyllaries in a single series, mostly of about equal lengths, but heads often with some shorter phyllaries at base.

  c. Bracteoles present at base of involucre, and black-tipped........... ...............................................55. *Senecio*

  cc. Bracteoles, if present, not black-tipped.

  d. Disk flowers purple ...........................52. *Petasites*

  dd. Disk flowers yellow, white, or orangish, never purple.

  e. Blades triangular or hastate, margins serrate or dentate .......... ...............................................54. *Cacalia*

  ee. Blades not as above.

  f. Blades palmately veined; phyllaries 4 or 5 ...........54. *Cacalia*

  ff. Blades not palmately veined; phyllaries more than 5.

  g. Plants annual; the outer flowers carpellate; corolla filiform with limb 2–4-toothed; involucre about 1.5 mm high, cylindric ...........................................53. *Erechtites*

  gg. Plants perennial; flowers perfect; corollas tubular, 5-cleft ...............................................54. *Cacalia*

  bb. Phyllaries in several series of unequal lengths ...........54. *Cacalia*

aa. Heads radiate.

  b. Flowers appearing in early spring before the leaves; heads solitary on scaly scapes; leaves angular but unlobed, white-tomentose on lower surface ....................................51. *Tussilago*

  bb. Flowers appearing after leaves; cauline leaves present and usually pinnatifid ...........................................55. *Senecio*

## 51. Tussilago L. COLTSFOOT

Heads radiate, solitary on white-woolly, scaly stems. Phyllaries in a single series, equal. Receptacle flat, naked. Ray flowers numerous, carpellate, seed-forming. Disk flowers apparently perfect but not seed-forming. Pappus of numerous capillary bristles. Achenes linear, ribbed. Heads appearing singly in spring on scaly scapes before basal leaves develop. Perennial herbs. The genus contains but a single species, native to northern Europe and Asia. For further treatment of genus see Morton (1978).

1. TUSSILAGO FARFARA L. COLTSFOOT

Well-established in eastern Ohio, where it occurs in damp soils, flowering early in the spring, on scaly scapes from horizontal rhizomes. Later the basal leaves develop which are long-petiolate, cordate, shallowly lobed, glabrous above, and white-tomentose beneath. Supposedly the plant has been ground and used as a cough remedy. March–early May.

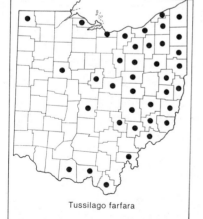

Tussilago farfara

× ½

Tussilago farfara

### 52. Petasites Mill. SWEET COLTSFOOT

Heads discoid (in ours), somewhat dioecious, in an elongated racemiform arrangement. Basal leaves large, long-petiolate, reniform or round-cordate or merely dentate-angled, base of petiole sheathing and dilated. Achenes 5–10-ribbed. Pappus of capillary bristles. Perennials.

1. PETASITES HYBRIDUS (L.) Gaertn., Mey. & Scherb. BUTTERFLY DOCK. BUTTERBUR

Introduced from Europe. Rare, known only from Mahoning County (not illustrated or mapped). April–June.

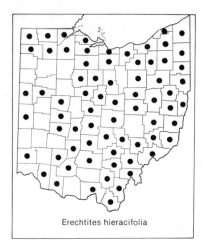

Erechtites hieracifolia

### 53. Erechtites Raf. FIREWEED

Heads discoid. Phyllaries equal, linear, herbaceous with attenuate tips. Receptacle flat, naked. Flowers dull white to creamy, the outer carpellate and seed-forming, the inner appearing perfect but rarely setting seed. Pappus capillary, of soft, white, slender bristles. Annual herbs with lanceolate to ovate-lanceolate, serrate to irregularly toothed or lobed leaves.

**1. Erechtites hieracifolia** (L.) Raf. FIREWEED. PILEWORT

Native to America north of Mexico, where it grows in a wide variety of habitats, but mainly in marshes, woodlands, and on lands recently burned, hence the name "fireweed". Aug.–Oct.

R. W. Long and S. J. Wood (unpublished information on herbarium specimens) determined that many of our specimens with reduced upper leaves are assignable to var. *intermedia* Fern. However, differences between those specimens and others assignable to var. *hieracifolia* appear to be due mainly to environmental stress. For treatments of the genus, see Belcher (1956) and Barkley and Cronquist (1978).

### 54. Cacalia L. INDIAN-PLANTAIN

Heads discoid, 5–many-flowered, arranged in flat corymbs. Receptacle essentially flat (sometimes with raised central portion), naked. Flowers perfect; corollas deeply 5-cleft, white, pinkish, or occasionally tawny. Pappus of soft capillary bristles. Achenes nerved and smooth, cylindric. Tall, alternate-leaved herbs, often with milky juice. For a taxonomic study of the genus, see Pippen (1978).

a. Leaves pinnately veined, the lower and larger ones triangular-hastate . . .
. . . . . . . . . . . . . . . . . . . . . . . . . . . . . . . . . . . . . . . . . . . . .1. *C. suaveolens*
aa. Leaves either palmately veined, or entire and strongly 5–7-nerved.
 b. Leaves entire or nearly so, strongly 5–7-nerved. . . . . . .2. *C. plantaginea*
 bb. Leaves palmately veined, shallowly lobed or toothed.
  c. Stems grooved longitudinally; leaves green, not glaucous. . . . . . . . . . . .
. . . . . . . . . . . . . . . . . . . . . . . . . . . . . . . . . . . . . . .3. *C. muhlenbergii*
  cc. Stems smooth or nearly so, glaucous; leaves glaucous beneath . . . . . . .
. . . . . . . . . . . . . . . . . . . . . . . . . . . . . . . . . . . . . . . .4. *C. atriplicifolia*

**1. Cacalia suaveolens** L. SWEET-SMELLING INDIAN-PLANTAIN

Fibrous-rooted perennial; stems glaucous, grooved longitudinally; leaves triangular-hastate, pinnately veined, long-tapering, sharply serrate-dentate, petiolate, the upper leaves less hastate; inflorescence corymbose; receptacle flat, pitted; phyllaries linear, attenuate. Woodland borders, thickets, and low ground. July–Oct.

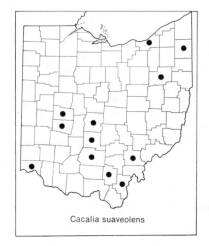

Cacalia suaveolens

×1

× ½

**Erechtites hieracifolia**

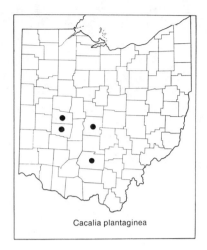

Cacalia plantaginea

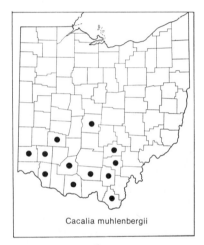

Cacalia muhlenbergii

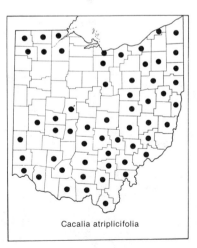

Cacalia atriplicifolia

2. **Cacalia plantaginea** (Raf.) Shinners

   *C. tuberosa* Nutt.

Stems angled and grooved, emerging from a thick tuberous root; leaves 5–7-nerved, thick and green, the lower leaves essentially entire, ovate, tapering to long petioles, the upper with short petioles, often toothed at apex. Marly bogs, wet prairies, and damp fields. June–Aug.

3. **Cacalia muhlenbergii** (Sch. Bip.) Fern.

   *C. reniformis* Muhl., not Lam.

Fibrous-rooted perennial; stems to 3 m tall, glabrous, obviously grooved; leaves palmately veined, glabrous, green above and beneath, often shallowly lobed, the lower leaves reniform and long-petiolate, the upper reduced, ovate and short-petiolate; heads many in a somewhat flat-topped arrangement. Woodlands and woodland borders. June–Sept.

4. **Cacalia atriplicifolia** L.   PALE INDIAN-PLANTAIN

Similar to *C. muhlenbergii* but stem glaucous and only slightly striate; leaves palmately veined and angulate-lobed, glaucous beneath, toothed. Thin woodlands, woodland borders and openings. Common. July–Sept.

55. Senecio L.   RAGWORT. GROUNDSEL. SQUAW-WEED

Heads radiate or discoid. Phyllaries herbaceous, equal, in a single series, small bracteoles often present. Receptacle naked, flat, honeycombed. Ray flowers carpellate, seed-forming, yellow, orangish or absent. Disk flowers perfect, seed-forming, yellow. Pappus white, soft-capillary. Achenes terete, nerved. Annual or perennial herbs with alternate, entire to variously toothed or divided leaves.

A very large genus with world-wide distribution. Many species exhibit a great deal of variation which appears to be the result of hybridization. Eight species can be recognized in Ohio. For further study see Barkley (1978).

a.  Plants leafy to the summit, the leaves not greatly reduced upward.

  b.  Heads discoid or appearing so.

    c.  Ray flowers absent; bracteoles below phyllaries black-tipped . . . . . . . . . . . . . . . . . . . . . . . . . . . . . . . . . . . . . . . . . . . . . . . . . . . . . . . . . . . . . .1. *S. vulgaris*

    cc.  Ray flowers present but short, heads appearing rayless; bracteoles below phyllaries not black-tipped . . . . . . . . . . . . . . . . . . . . . . .2. *S. sylvaticus*

  bb.  Heads radiate; leaves mostly basal, lyrate-pinnatifid, the terminal lobe large, rounded . . . . . . . . . . . . . . . . . . . . . . . . . . . . . . . . . .3. *S. glabellus*

aa.  Plants with leaves rapidly reduced upward, basal and rosette leaves undivided, the upper mostly pinnatifid.

  b.  Plants usually persistently tomentose on stem, lower leaf surfaces, and phyllaries. . . . . . . . . . . . . . . . . . . . . . . . . . . . . . . . . . . . . . . . . . . .4. *S. plattensis*

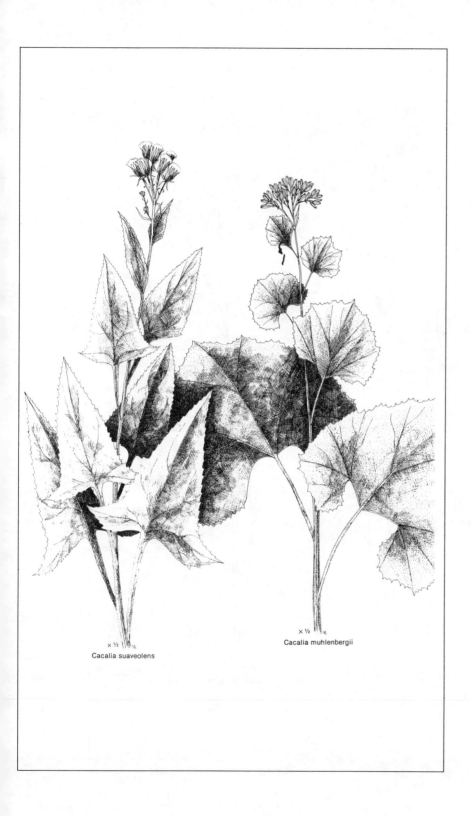

× ½

Cacalia suaveolens

× ½

Cacalia muhlenbergii

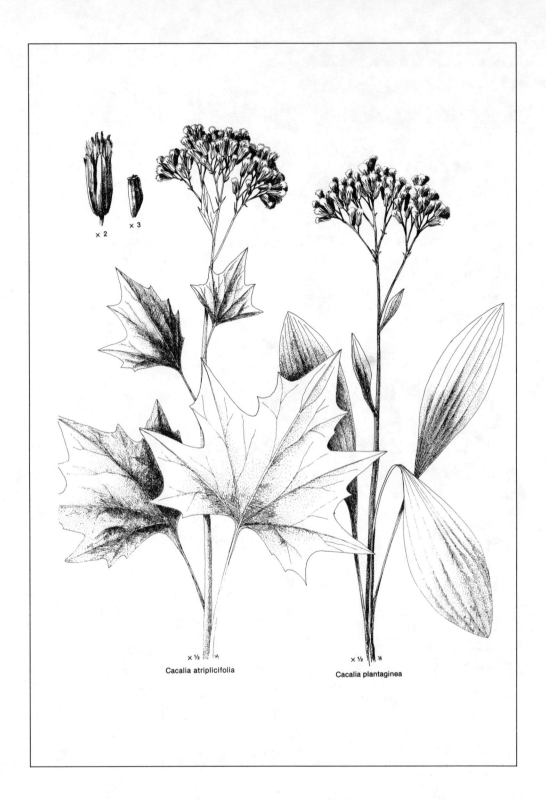

× 2      × 3

× ½      Cacalia atriplicifolia

× ½      Cacalia plantaginea

bb. Plants not tomentose except often in leaf axils and at base of stem.

   c. Basal leaves ovate to elliptic, cordate at base; stolons not well-developed . . . . . . . . . . . . . . . . . . . . . . . . . . . . . . . . . . . . . . .5. *S. aureus*

   cc. Basal leaves mostly oblanceolate, if elliptic, then tapering at base.

     d. Plants strongly stoloniferous; leaves, if elliptic, tapering . . . . . . . . . . .
. . . . . . . . . . . . . . . . . . . . . . . . . . . . . . . . . . . .6. *S. obovatus*

     dd. Plants not strongly stoloniferous, although a few short stolons sometimes present.

       e. Heads few, rarely as many as 20; stems mostly glabrous . . . . . . . . . . .
. . . . . . . . . . . . . . . . . . . . . . . . . . . . . . . . . . .7. *S. pauperculus*

       ee. Heads usually more than 20; stems densely woolly at base, elsewhere soon glabrous . . . . . . . . . . . . . . . . . . . . . . . . . . . . .8. *S. anonymus*

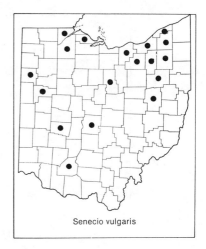

Senecio vulgaris

1. SENECIO VULGARIS L.   COMMON GROUNDSEL (of Europe)

Plants annual; stems to 0.5 m tall, leafy, mostly glabrous; leaves coarse, fleshy, irregularly toothed, usually pinnatifid, the upper sessile and clasping, the lower tapering to base; heads many, discoid; flowers perfect, fertile; phyllaries glabrous, linear, subtended by short, black-tipped bracteoles; pappus bristles numerous, slightly longer than corollas; achenes pubescent on nerves or angles. A native of Europe, weedy in Ohio and sometimes becoming a pest, with scattered distribution in cultivated fields and waste ground. May–early fall.

2. SENECIO SYLVATICUS L.

Plants annual; stems short, sparingly pubescent, leafy; leaves sparingly pubescent, sometimes viscid-glandular, pinnatifid, irregularly toothed; heads several; rays short and inconspicuous; bracteoles below phyllaries usually present but not black-tipped; pappus bristles numerous, slightly exceeding the disk; achenes minutely pubescent. Native to Europe, where it is a troublesome weed. Although common in the eastern part of the United States, this species has been collected only twice in Ohio, both 1892 specimens in The Ohio State University Herbarium. It is possibly often overlooked, being mistaken for *S. vulgaris*. July–Sept.

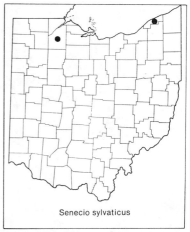

Senecio sylvaticus

3. **Senecio glabellus** Poir.   BUTTERWEED. YELLOWTOP

Plants annual, fibrous-rooted; stems soft, glabrous, hollow, to 1 m tall, usually solitary and unbranched to the inflorescence; leaves glabrous, only slightly reduced upward, pinnatifid, the lobes usually rounded; heads several; rays showy, lemon-yellow; phyllaries linear-lanceolate, bracteolate; achenes minutely pubescent. Uncommon in Ohio, it occurs at damp woodland borders, and in swamps and wet ditches. Midspring–July.

Senecio glabellus

× ½
Senecio plattensis

× ½
Senecio vulgaris

× ½
Senecio glabellus

### 4. **Senecio plattensis** Nutt.

Plants perennial, sometimes becoming stoloniferous; stems to 0.5 m tall, usually several from forking crown, tomentose-lanate throughout; leaves tomentose above and beneath, basal leaves narrow to ovate or broadly elliptic, crenate-serrate, tapering to petiole, usually unlobed, cauline leaves reduced upward, sessile, pinnatifid or not; heads many, rays showy; phyllaries linear-lanceolate; achenes hirtellous on angles. Calcareous rocks and old fields. Uncommon. May–July.

### 5. **Senecio aureus** L.   Golden Ragwort

Plants perennial, fibrous-rooted, and with short stolons; stems to 1 m tall, glabrous when mature; basal leaves long-petiolate, usually broad and cordate, rounded at apex, crenate, occasionally serrate, cauline leaves nearly always pinnatifid, becoming sessile and reduced upward; heads many; phyllaries glabrous, linear, often purple-tipped; achenes glabrous. Common in moist woodlands and low places. April–July.

Many varieties have been described within this quite variable species, but I fail to see their validity, at least among Ohio plants. For a taxonomic study of the species, see Barkley (1962).

Senecio plattensis

Senecio aureus

Senecio sylvaticus

× 1
Senecio pauperculus

× ½
Senecio aureus

× ½
Senecio obovatus

6. **Senecio obovatus** Muhl. ex Willd.   Roundleaf Squaw-weed

Plants perennial with well-developed stolon system, basal rosettes usually present; stems glabrous when mature, to 1 m tall; basal leaves broadly elliptic or ovate or broadly oblanceolate, tapering, not cordate at base, crenate or crenate-serrate; cauline leaves reduced upward, becoming sessile, pinnatifid; heads numerous, usually on long slender peduncles; phyllaries abruptly acuminate; achenes hirtellous on angles. Common in open, dry, rich woods. April–July.

7. **Senecio pauperculus** Michx.   Balsam Squaw-weed

Plants perennial, fibrous-rooted; stems short, rarely 20 cm tall, soon glabrate, except in leaf axils; basal leaves mostly oblanceolate, rounded at the apex, tapering at base onto petiole, crenate to subentire; cauline leaves reduced upward, becoming sessile, pinnatifid; heads few, 10–15; phyllaries linear, greenish, tapering at apex; ray flowers deep yellow, attractive; achenes glabrous. Usually occurring on calcareous soils, especially along cliffs and rocky shores of Lake Erie and in a few southern counties. May–July.

8. Senecio anonymus A. Wood

   *S. smallii* Britt.

Plants perennial; fibrous-rooted; stems 1–several, soon glabrous except persistently woolly at base; basal leaves long, spatulate-oblanceolate to lance-obovate, tapering to petiole, crenate or serrate; cauline leaves pinnatifid or pectinate-pinnatifid, becoming sessile upward; heads numerous on slender peduncles; phyllaries linear, sometimes suffused with purple; achenes hispidulous. Long (1959) suggests this species may be a recent migrant to Ohio. Scattered distribution in open fields and pastures, and along roadsides. May–July.

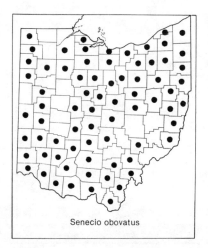

Senecio obovatus

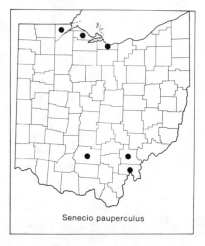

Senecio pauperculus

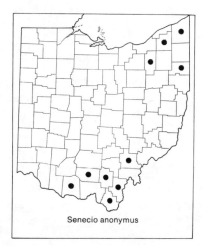

Senecio anonymus

Tribe VIII. Senecioneae   205

× ½

Senecio anonymus

# TRIBE IX. CYNAREAE SPRENG.

a. Flowers of head uniform; achenes attached to receptacle by base, or (in *Centaurea*) attached obliquely.

  b. Receptacles densely bristly-hairy.

    c. Phyllaries rigid, hooked at tip; leaves not prickly or spiny . . . . . . . . . . . . . . . . . . . . . . . . . . . . . . . . . . . . . . . . . . . . . . . . . . . . . . . . . . . . . . . . . . . . . . . . .56. *Arctium*

   cc. Phyllaries not hooked at tip.

    d. Pappus bristles plumose; plants spiny.

      e. Filaments separate . . . . . . . . . . . . . . . . . . . . . . . . . . . . . .58. *Cirsium*

     ee. Filaments united, at least below . . . . . . . . . . . . . . . . . . .60. *Silybum*

    dd. Pappus bristles naked, not plumose, merely rough or denticulate or setose, or pappus none.

      e. Phyllaries tipped with appendages, or rarely tipped with long spines; stems and leaves without spines; achenes obliquely attached to receptacle . . . . . . . . . . . . . . . . . . . . . . . . . . . . . . . . . . . . . . . . . . . . .61. *Centaurea*

     ee. Phyllaries linear to ovate, tipped with only short spines; wing of stems and leaves spine-toothed; achenes attached to receptacle by base . . . . . . . . . . . . . . . . . . . . . . . . . . . . . . . . . . . . . . . . . . . . .57. *Carduus*

  bb. Receptacles naked or with a few bristles, deeply honeycombed; stem broadly winged, teeth of wing and leaf blades spine-tipped; pappus barbellate; plants white-woolly . . . . . . . . . . . . . . . . . . . . . . .59. *Onopordum*

aa. Flowers of head often enlarged, especially marginal ones, and often appearing ray-like; achenes attached obliquely at base to receptacle . . . . . . . . . . . . . . . . . . . . . . . . . . . . . . . . . . . . . . . . . . . . . . . . . . . . . . . . . . . .61. *Centaurea*

## 56. Arctium L. BURDOCK

Heads discoid, globose to subglobose. Phyllaries multiseriate, stiff, tipped with inwardly hooked spines. Receptacle flat, bristly. Flowers perfect, seed-forming, corollas pink, white, or purple. Pappus of scabrous, deciduous bristles. Achenes many-nerved, glabrous, truncate. Biennial herbs with very large, ovate, cordate, alternate, mostly entire leaves.

a. Inflorescence corymbose, heads long-pedunculate . . . . . . . . . .1. *A. lappa*

aa. Inflorescence racemose, heads sessile or short-pedunculate . . . . . . . . . . . . . . . . . . . . . . . . . . . . . . . . . . . . . . . . . . . . . . . . . . . . . . . . . . . . . . . . . . .2. *A. minus*

## 1. ARCTIUM LAPPA L. GREAT BURDOCK

Stems tall, to 1.5 m or more; leaves petiolate, reduced upward, petioles mostly solid, blades broadly ovate, large, cordate, short-tomentose beneath, essentially glabrous above; inflorescence corymbose, peduncles to 10 cm or longer, occasionally glandular-hairy; heads 2.5–4.0 cm wide; phyllaries greenish, long,

×1

Arctium minus

×1

Arctium lappa

narrow, hooked at tip. Native to Eurasia, with a scattered distribution in Ohio, in moist waste places. Not so common as *A. minus*, and apparently capable of hybridizing with it. July–late fall.

2. ARCTIUM MINUS Schkuhr.    COMMON BURDOCK

Stems to 1.5 m or more tall; leaves petiolate, reduced upward, petioles mostly hollow, especially the lower, blades broadly ovate, cordate, short-tomentose to glabrate beneath, essentially glabrous above; inflorescence racemose; heads short-pedunculate to subsessile, 1.5–2.5 cm wide, peduncles short-glandular to glabrous; phyllaries greenish, hooked at apex, glabrous or often cobwebby. Naturalized from Eurasia, and well-established as a weed in waste places, especially around old farm buildings. July–late fall.

A similar species, *A. tomentosum* Mill., more common in New England and along the Atlantic coast, has been reported from Ohio, and some specimens in The Ohio State University Herbarium are so identified, but in my opinion they are within the limits of variation of *A. minus*.

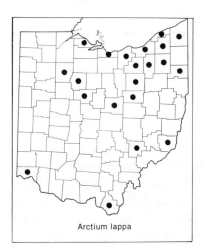

Arctium lappa

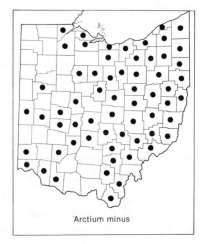

Arctium minus

## 57. Carduus L.  PLUMELESS THISTLE

Heads large, discoid. Phyllaries spine-tipped, often reflexed with age. Receptacle flat or convex, bristly. Flowers purple, pink, or white, tube slender with long, narrow lobes. Pappus of many smooth or merely rough capillary bristles (which is the best character for separating this genus from *Cirsium*, which has plumose bristles). Annual or mostly biennial coarse spiny herbs native to North Africa, Europe, and Asia. Our species are introduced, but fairly well-established along roadsides and in waste places.

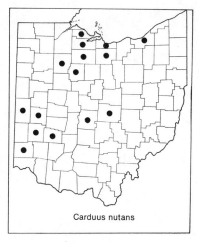

Carduus nutans

Carduus acanthoides

a.  Heads solitary and nodding, showy, to 7 cm wide; phyllaries 4 to 8 mm wide . . . . . . . . . . . . . . . . . . . . . . . . . . . . . . . . . . . . . . . . . . .1. *C. nutans*

aa. Heads often clustered, upright, less than 3 cm wide; phyllaries less than 2 mm wide . . . . . . . . . . . . . . . . . . . . . . . . . . . . . . .2. *C. acanthoides*

### 1. CARDUUS NUTANS L.   NODDING THISTLE. MUSK-THISTLE

Biennial, to 1 m tall; stems spiny, generally winged by decurrent leaf bases; leaves glabrous, sometimes sparingly villous on veins beneath, deeply lobed; heads most often solitary and nodding at the ends of branches, usually large, to 7 cm wide; phyllaries broad, 4–8 mm wide, long, the outer often reflexed, their tips spiny, the inner often soft, spineless. Native to Eurasia, and widely scattered in Ohio, found along roadsides and in waste places. Stuckey and Forsyth (1971) demonstrated a dramatic correlation between dolomite or limestone bedrock highs and the distribution of this species. July–Oct.

### 2. CARDUUS ACANTHOIDES L.

Biennial, averaging about 0.5 m tall; stems stout and even more strongly spiny than those of *C. nutans*; leaves deeply lobed or pinnatifid, sparingly villous beneath, especially on midveins; heads erect and usually clustered at the ends of branches, small and less than 3 cm wide; phyllaries narrow, about 2 mm wide, spreading and mostly erect, the outer tipped with spines, the inner spineless. Native to Europe; uncommon in Ohio, where it occurs in waste places and along roadsides. July–Oct.

### 58. Cirsium Mill.   THISTLE

Heads discoid, large. Phyllaries multiseriate, imbricate, usually spine-tipped. Receptacle flat to slightly convex, bristly. Flowers perfect or sometimes imperfect, pink, purple, or white, corolla tubes slender with long, narrow lobes. Pappus of plumose bristles, united at base and falling as a ring. Achenes usually nerved, flattened or nearly quadrangular. Coarse and spiny perennials and biennials with alternate usually pinnatifid leaves. A genus of many species, perhaps 200, native to North America, Europe, and Asia. For further information on the genus, see Moore and Frankton (1974).

a.  Leaves obviously decurrent on the stem as spiny wings. . . . .1. *C. vulgare*

aa. Leaves not decurrent on the stem or only slightly so.

b.  Heads small, 1–3 cm high (as pressed); flowers imperfect, species imperfectly dioecious . . . . . . . . . . . . . . . . . . . . . . . . . . . . . . .2. *C. arvense*

bb. Heads larger than 2 cm high (as pressed).

c.  Leaves densely white-tomentose beneath, glabrous or hispid above.

d.  Heads subtended by leafy bracts, peduncles short and leafy; heads 2.0–3.5 cm high.

e.  Cauline leaves deeply pinnatifid, the lobes linear-lanceolate . . . . . . . . . . . . . . . . . . . . . . . . . . . . . . . . . . . . . . . . . . . . .3. *C. discolor*

× 2

× ½                    Carduus nutans

× ½          Carduus acanthoides

ee. Cauline leaves entire, shallowly toothed, or very shallowly lobed.
. . . . . . . . . . . . . . . . . . . . . . . . . . . . . . . . . . . . . . . . . . . . . .4. *C. altissimum*
dd. Heads naked at base or bracts few and reduced; heads usually 2.0–2.5 cm high, peduncles long. . . . . . . . . . . . . . . . . . . . .5. *C. carolinianum*
cc. Leaves glabrous or pubescent, if tomentose only slightly so, usually greenish on both surfaces.
d. Phyllaries not spine-tipped (often mucronate with glutinous band); involucre cobwebby . . . . . . . . . . . . . . . . . . . . . . . . . . . .6. *C. muticum*
dd. Phyllaries spine-tipped.
e. Outer phyllaries with terminal prickle 4–6 mm long, back of phyllary rarely with glutinous band; root solid; biennial . . . . . .7. *C. pumilum*
ee. Outer phyllaries with short terminal prickle 1.5–3.0 mm long, back of phyllary with glutinous band; root hollow; perennial . . . . . . . . . .
. . . . . . . . . . . . . . . . . . . . . . . . . . . . . . . . . . . . . . . . . . . . .8. *C. hillii*

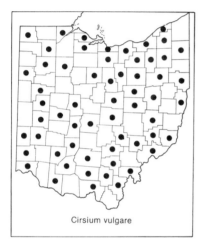

Cirsium vulgare

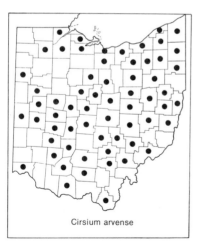

Cirsium arvense

1. Cirsium vulgare (Savi) Tenore   Common Thistle. Bull Thistle
*C. lanceolatum* Scop., not Hill
Biennial weed to 2 m tall; stems with spiny wings due to decurrent leaf bases, spreading-hirsute; leaves pinnatifid, pale-woolly beneath, hirtellous and green above, lobes armed with wicked, long, sharp-pointed spines; heads solitary or few at tips of prickly-winged branches; phyllaries attenuate and tipped with sharp, rigid prickles; flowers purple. Native to Eurasia; well-established in Ohio, where it occurs in open fields, pastures, waste places, and along road-sides. June–Oct.

2. Cirsium arvense (L.) Scop.   Canada Thistle
*Carduus arvensis* (L.) Robson
Perennial, rhizomes abundant, to 5 m tall; stems leafy, essentially glabrous; leaves rigid, short-woolly on both surfaces, more so beneath, oblong to lanceo-late, not extensively pinnatifid but rather sinuate-pinnatifid, very prickly mar-gined, mostly sessile; heads ovoid-cylindric, numerous, often clustered; phyllaries mostly with subulate tips, the inner with lance-attenuate tips; flowers pinkish-purple or occasionally white. Native to Eurasia and now a ubiquitous weed of fields and waste places. The species spreads rapidly and is difficult to eradicate because of extensive and deep rhizomes. June–Oct.

Two varieties can readily be recognized in the state. Variety *horridum* Wimm. & Grab. is the most common and can be recognized by its sinuate-pinnatifid, spiny leaves. Variety *arvense* (var. *mite* Wimm. & Grab., var. *integri-folium* Wimm & Grab., var. *vestitum* Wimm. & Grab.) has merely toothed and weakly-spined leaves with very shallow lobes, and is less common. There ap-pear to be no ecological or distributional differences between the varieties.

× ½
Cirsium discolor

× ½
Cirsium vulgare

× ½
Cirsium arvense

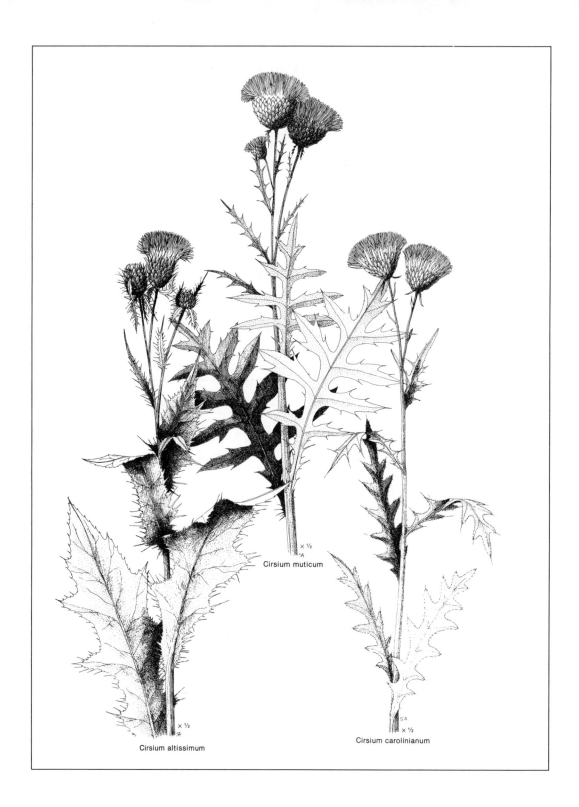

Cirsium muticum × ½

Cirsium altissimum × ½

Cirsium carolinianum × ½

3. **Cirsium discolor** (Muhl.) Spreng.

*Carduus discolor* (Muhl.) Nutt.

Biennial or perennial; stems furrowed, hirtellous or glabrate, 1–3 m tall, branching near summit; leaves divided nearly to the midrib into linear-lanceolate to narrowly-oblong, bristly lobes, white-felty beneath, green and essentially glabrous above, the upper leaves reduced and sessile, forming a false involucre below the head; heads solitary on leafy branches, 2.0–3.5 cm high; phyllaries with weak, spreading spines; flowers purple or occasionally white. For further information on morphological variation see Davidson (1963). Fields, waste places, and woodland borders. Common. July–Oct.

4. **Cirsium altissimum** (L.) Spreng.   TALL THISTLE

*Carduus altissimus* L.

Perennial, often taller than C. *discolor*; stems hirsute to subglabrate; leaves mostly undivided, ovate-lanceolate, sinuate-toothed or shallowly lobed, densely white-tomentose beneath, scabrous-hirsute to subglabrate above, spiny-toothed on margins; heads solitary to several, 2.0–3.5 cm high, on leafy peduncles; outer phyllaries tipped with spines, 2–5 mm long; flowers purple, occasionally white. Fields, waste places, and frequently on flood plains. July–Oct.

See Ownbey (1964) for information on hybridization between this species and C. *discolor*.

5. **Cirsium carolinianum** (Walt.) Fern. & Schub.   CAROLINA THISTLE

*Carduus carolinianus* Walt.

Biennial with fibrous roots; stems glabrous above, often arachnoid below; leaves white-tomentose beneath, glabrous or merely hirsute above, the lower leaves somewhat irregularly pinnatifid, the upper reduced and mostly unlobed

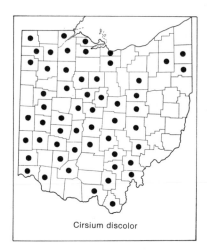

Cirsium discolor

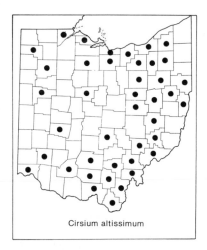

Cirsium altissimum

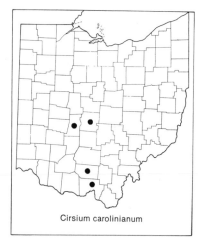

Cirsium carolinianum

and spinulose-margined; heads solitary or few on essentially naked peduncles, the peduncles occasionally with a few scattered linear bracts; phyllaries with short, slender spine-tips; flowers magenta-purple. Uncommon in open woodlands and dry open habitats in the southern part of Ohio. May–June.

6. **Cirsium muticum** Michx.   Swamp Thistle

Biennial, 1 m or less tall; stems glabrous above, hirtellous below, hollow below; leaves deeply pinnatifid, the segments weakly spiny, subscabrous above, sparingly hirsute beneath; leaves reduced above; heads usually several on long peduncles, spineless, conspicuously cobwebby; flowers purple. Swamps, bogs, and similar wet habitats. July–Oct.

7. **Cirsium pumilum** (Nutt.) Spreng.   Pasture Thistle

*Cirsium odoratum* (Muhl.) Petrak

*Carduus pumilus* Nutt.

Biennial to 0.5 m tall; stems solid, hairy, sparingly branched; leaves lobed to pinnatifid, lanceolate-oblong, green, very prickly on margins and lobes; heads large, short-pedunculate, with or without leafy bracts; phyllaries with mostly erect spine-tips, usually without glutinous ridge on back; flowers purple or occasionally white. Pastures, open fields, and occasionally in open woods. Uncommon. July–Sept.

8. **Cirsium hillii** (Canby) Fern.   Hill's Thistle

Similar to C. *pumilum* above, but differing chiefly in having phyllaries with a distinct dark glutinous dorsal ridge. Rare. Sandy, open fields. June–Aug.

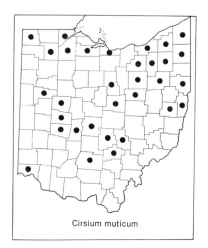

Cirsium muticum

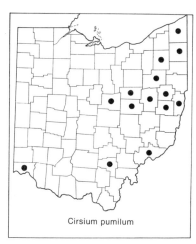

Cirsium pumilum

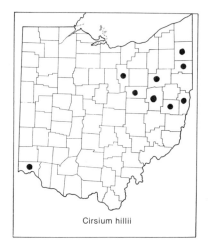

Cirsium hillii

Cirsium pumilum

× ½

Cirsium hillii

× ½

### 59. Onopordum L. SCOTCH THISTLE. COTTON-THISTLE

Heads discoid, large, solitary at ends of branches. Phyllaries imbricate, spiny on margins, strongly spiny on tips. Receptacle flat, honeycombed, scarcely bristly. Flowers perfect, seed-forming, purplish-pink to white, slender, the lobes long and narrow. Pappus reddish or brownish, with plumose or short-barbellate bristles. Achenes glabrous, truncate at apex, oblong or quadrangular, often somewhat compressed. Introduced and occasionally naturalized biennial, white-tomentose or woolly, spiny herbs with alternate, decurrent, dentate to pinnatifid leaves. Perhaps 40 species of Europe, North Africa, and Asia.

### 1. ONOPORDUM ACANTHIUM L. SCOTCH THISTLE

Naturalized in Ohio, where it occurs in pastures and fields, and along roadsides. Fairly uncommon. July–Oct.

× ½
Onopordum acanthium

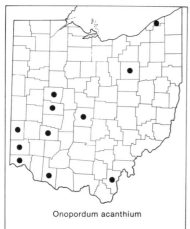

Onopordum acanthium

## 60. Silybum Adans. MILK-THISTLE

Heads discoid, flowers tubular and perfect. Phyllaries imbricate in about three series, spreading, reflexed at maturity, broad at base, spiny-margined and tipped with a stout spine. Receptacle flat, bristly. Corollas purple. Achenes glabrous, flattened. Pappus of slender bristles which fall as a ring. Tall biennials with large, white-mottled, clasping leaves that are decurrent onto the stem.

1. SILYBUM MARIANUM (L.) Gaertn.   MILK-THISTLE

Native to the Mediterranean region and well-established in northwestern United States, but rare in Ohio. Waste places. Early summer–July.

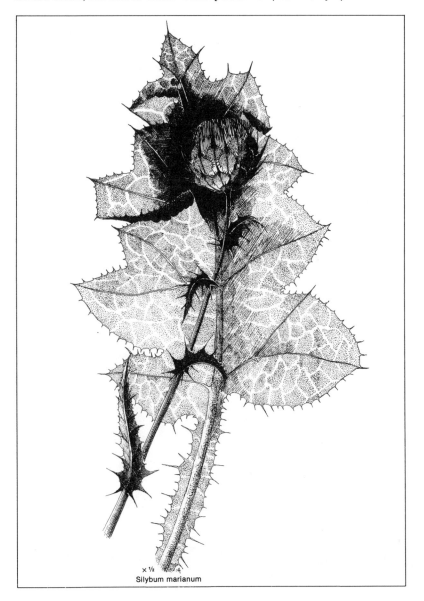

× ½
Silybum marianum

Silybum marianum

## 61. Centaurea L. STAR-THISTLE

Heads discoid. Phyllaries multiseriate, spiny or variously appendaged, serrulate, dentate or pectinate. Receptacle flat, densely bristly. Flowers pink, purple, or white, seeds forming in central flowers, marginal flowers neutral and often with enlarged corollas appearing ligulate. Pappus various, of bristles, scales, or reduced, or absent. Achenes obliquely attached to receptacle. Annual, biennial, or perennial herbs with alternate, sinuate, entire or pinnatifid leaves.

A genus of several hundred species of the Mediterranean region and a few native to North and South America. One species, *C. americana* Nutt., is native to the southern plains states and Mexico but not present in Ohio, except as a garden plant. Many other species have long been cultivated in gardens, and their seeds have scattered to roadsides and field margins. It seems best to regard all the species collected and represented in our herbaria as escapes.

a. Phyllaries armed with long spines . . . . . . . . . . . . . . . . . . .1. *C. solstitialis*
aa. Phyllaries not armed as above, entire, lacerate, or fringed.
  b. Plants evidently white-woolly; phyllaries with triangular and upright teeth . . . . . . . . . . . . . . . . . . . . . . . . . . . . . . . . . . . . . . . . .2. *C. cyanus*
  bb. Plants not white-woolly, at most close-tomentose; phyllaries various, but not with triangular and upright teeth.
    c. Heads large, 4.5–6.0 cm broad, 2.0–2.5 cm high; outer phyllaries with pectinate tips about 4 mm long. . . . . . . . . . . . . . . . . . .3. *C. scabiosa*
    cc. Heads less than 4.0 cm broad.
      d. Leaves deeply pinnatifid with narrow segments; outer phyllaries with pectinate tips about 2 mm long. . . . . . . . . . . . . . . . . .4. *C. maculosa*
      dd. Leaves unlobed, often denticulate, a few basal leaves occasionally pinnatifid, if so, then with broad segments.
        e. Phyllaries entire, serrulate, or irregularly denticulate.
          f. Plants bushy, much branched; pappus deciduous; appendages of inner phyllaries not bifid . . . . . . . . . . . . . . . . . . . . . . . . .5. *C. repens*
          ff. Plants sparsely branched; pappus persistent; appendages of inner phyllaries bifid. . . . . . . . . . . . . . . . . . . . . . . . . . . . . . .6. *C. jacea*
        ee. Outer phyllaries pectinate.
          f. Pectination only at the summit of the phyllary . . . . . . . .7. *C. dubia*
          ff. Pectination over at least half of the phyllary . . . . . . . . .8. *C. nigra*

1. CENTAUREA SOLSTITIALIS L. YELLOW STAR-THISTLE. BARNABY'S THISTLE
Annual or biennial; stems sparsely tomentose, winged; basal leaves usually lyrate-pinnatifid, the middle leaves linear and decurrent; heads several; outer phyllaries long spine-tipped, the inner hyaline; corollas yellow; pappus absent

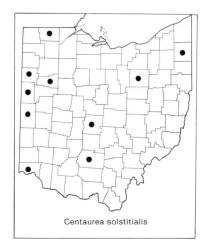

Centaurea solstitialis

× 2

Centaurea maculosa

× ½

Centaurea solstitialis

× ½

× 2

Centaurea cyanus

× ½

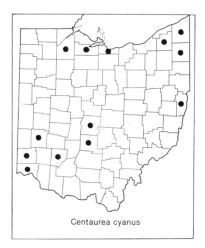

Centaurea cyanus

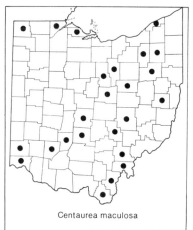

Centaurea maculosa

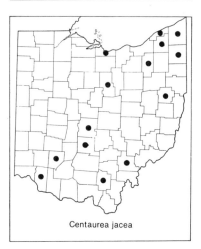

Centaurea jacea

in outer flowers. Native to the Mediterranean area and rapidly becoming naturalized in this country, especially in the west. At times a weedy species in fields and along roadsides. July–Sept.

2. CENTAUREA CYANUS L.    BACHELOR'S-BUTTON. CORNFLOWER

Annual; stems slender-branched, white-woolly; leaves entire, linear, the lower often denticulate or slightly lobed; phyllaries linear with upright and triangular teeth along margin; corollas blue, white, or rose, the marginal enlarged and attractive. A native of Europe often grown as an ornamental in borders and established as an escape in many areas along roadsides and in waste areas. May–Oct.

3. CENTAUREA SCABIOSA L.    SCABIOSA

Perennial; stems angled; leaves often dense at base and persisting from previous year, the lower pinnately divided, the upper reduced; heads 1–few, showy, 4.5–6.0 cm wide on elongated peduncles; phyllaries pectinate with darkened fringe; flowers rose to purple. Native to Europe; rare in Ohio, but apparently spreading to open fields and waste places in some areas of the northeastern counties (not illustrated or mapped). June–Sept.

4. CENTAUREA MACULOSA Lam.    SPOTTED KNAPWEED

Biennial; stems slender, wiry, to 1 m tall; leaves canescent, soon sparsely puberulent, punctate, deeply pinnatifid, lobes narrow; phyllaries striate, the outer and middle with short, pectinate tips; outer flowers enlarged, pinkish. A native of Europe, commonly established in waste places and along roadsides in our range. June–Oct.

5. CENTAUREA REPENS L.

Perennial, short, much-branched; stems canescent; leaves linear-lanceolate, entire, dentate, or slightly lobed below, obscurely glandular-punctate; heads abundant; phyllaries greenish, broad, striate, with broad, nearly entire tips, the inner serrulate; flowers rose to purple; the large plumose pappus soon deciduous. Native to Asia, probably introduced with alfalfa seed many years ago. Long (1959) reports the single collection from Clinton County. Rare (not illustrated or mapped). June–Oct.

6. CENTAUREA JACEA L.    BROWN KNAPWEED

Perennial; stems to 1 m tall, glabrous, usually sparingly branched; leaves lanceolate, entire or sinuate-toothed, basal leaves long-petiolate, the upper reduced, sessile, few; heads abundant; pappus persistent; phyllaries with well-developed appendages, the outer irregularly lacerate, the inner mostly bifid; marginal flowers enlarged. A native of Europe, now widely established in fields and along roadsides. June–Sept.

7. CENTAUREA DUBIA Suter    SHORT-FRINGED KNAPWEED

   C. vochinensis Bernh.

Perennial; stems harsh, rough-pubescent or glabrous; lower leaves petiolate,

× 3

× 3

× 3

× ½

Centaurea dubia

× ½

Centaurea jacea

× ½

Centaurea nigra

Tribe IX. Cynareae    223

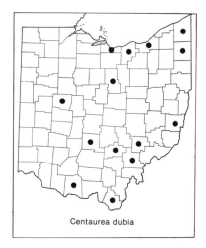

Centaurea dubia

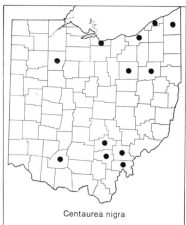

Centaurea nigra

oblanceolate, and sometimes lobed, the middle and upper reduced, sessile, toothed or entire; phyllaries pectinate only at summit, blackish; flowers rose-purple. Native to Europe; widely established in fields and waste places. July–Oct.

8. CENTAUREA NIGRA L.   BLACK KNAPWEED. SPANISH-BUTTONS

Perennial; stems harsh; leaves entire or sinuate-toothed, the basal lanceolate, petiolate, the upper reduced and sessile; phyllary appendages pectinate and well-developed, blackish, the outer and middle regularly pectinate over most of the margin, more extensively than in *C. dubia*; outer corollas usually absent. Native to Europe; widely established in fields and waste places. July–Oct.

## TRIBE X. LACTUCEAE CASS. (CICHORIEAE SPRENG.)

For literature pertaining to this tribe, see Vuilleumier (1973).

a.  Pappus absent; plants annual.
  b.  Cauline leaves present; peduncles not inflated . . . . . . . . . .62. *Lapsana*
  bb.  Leaves either all basal or cauline leaves reduced to mere bracts; peduncles inflated . . . . . . . . . . . . . . . . . . . . . . . . . . . . . . . . . . . . .63. *Arnoseris*
aa.  Pappus present; plants annual or perennial.
  b.  Pappus chaffy, or a mixture of chaff and bristles.
    c.  Involucre of two series, the inner of 8–10 phyllaries, the outer of 5 short phyllaries; pappus of uniform narrow scales; flowers blue or rarely pink or white . . . . . . . . . . . . . . . . . . . . . . . . . . . . . . . . . .64. *Cichorium*
    cc.  Involucre of a single series; pappus of scales alternating with bristles; flowers yellow or orange . . . . . . . . . . . . . . . . . . . . . . . . . . . . .65. *Krigia*
  bb.  Pappus plumose or capillary, rarely also with a few scales.
    c.  Pappus bristles of most flowers plumose or barbellate.
      d.  Leaf blades linear-lanceolate, entire . . . . . . . . . . . . . .69. *Tragopogon*
      dd.  Leaf blades toothed or lobed.
        e.  Cauline leaves present . . . . . . . . . . . . . . . . . . . . . . . . . . .68. *Picris*
        ee.  Leaves all basal or crowded near the base of stem.
          f.  Receptacles naked (or with a few hairs) . . . . . . . . . . .67. *Leontodon*
          ff.  Receptacles chaffy . . . . . . . . . . . . . . . . . . . . . . . . .66. *Hypochaeris*
    cc.  Pappus simple, not plumose, some white scales occasionally present.
      d.  Achenes fusiform with sharp points and with slender beaks; herbs scapose, the scapes hollow; heads solitary . . . . . . . . . . .70. *Taraxacum*
      dd.  Achenes without sharp points (not muricate).
        e.  Achenes flat or flattish.
          f.  Achenes with a beak or thickened summit; flowers fewer than 50 per head; involucre cylindric or conic; flowers blue, purple, or yellow . . . . . . . . . . . . . . . . . . . . . . . . . . . . . . . . . . . . . . . . . . . . . .72. *Lactuca*

ff. Achenes without beak; flowers more than 50 per head; involucre ovoid or campanulate; flowers yellow . . . . . . . . . . . . . . . .71. *Sonchus*

ee. Achenes columnar, not or only slightly flattened.

   f. Corollas yellow or orange.

      g. Achenes markedly narrowed at apex; phyllaries in two series, the outer much reduced; pappus white . . . . . . . . . . . . . . . .73. *Crepis*

      gg. Achenes truncate or only slightly narrowed at apex; phyllaries in three series; pappus gray, tawny, or brown . . . . . . . .75. *Hieracium*

   ff. Corollas white, cream, or pink; heads drooping . . . . .74. *Prenanthes*

## 62. Lapsana L. NIPPLEWORT

Heads ligulate, several to many in a panicle or corymb. Phyllaries biseriate, the outer bract-like and small, the inner about 8, longer, nearly equal. Corollas yellow. Pappus none. Achenes terete, nerved, narrowed at both ends. Slender branching annual herbs, glabrous to hirsute, with alternate ovate, entire to lyrate leaves. A genus of about nine species native to temperate Eurasia.

1. LAPSANA COMMUNIS L.   NIPPLEWORT

Introduced from Europe, a weed in some northern areas and apparently spreading southward in Ohio. Other than those from northern counties, the only additional collections are from Franklin and Delaware Counties. Supposedly, this species is strictly self-pollinating. Fields, gardens, and waste places. June–Oct.

Lapsana communis

## 63. Arnoseris Gaertn.   DWARF NIPPLEWORT

Heads ligulate, solitary at the ends of branches. Phyllaries uniseriate, equal, becoming slightly enlarged at maturity. Flowers perfect, seed-forming, corollas yellow. Pappus absent. Achenes ribbed, obovoid. Small glabrous or puberulent annual with basal, toothed and ovate-lanceolate to oblanceolate leaves, and with peduncles slightly inflated at maturity.

1. ARNOSERIS MINIMA (L.) Schweigg. & Koerte

Native to Europe, but rare in Ohio, where it occurs in lawns and waste places. Possibly it is overlooked because of its resemblance to the common dandelion. Early summer–fall.

## 64. Cichorium L.   CHICORY

Heads several-flowered, approaching 3.0 cm in diameter, sessile or nearly so, often 1–3 in the axils of reduced leaves. Phyllaries biseriate, the outer row shorter and usually spreading. Receptacle essentially naked and somewhat fim-

Arnoseris minima

brillate. Flowers showy, perfect, all ligulate, corollas blue to infrequently pink or white. Pappus of many narrow scales or a mere crown. Achenes 2–3 mm long, quadrangular, almost black, somewhat striate. Deep-rooted perennials; stems usually glabrous, to 1.5 m tall. Leaves various, the basal and cauline lanceolate, entire to pinnatifid, often runcinate, alternate, the cauline clasping or nearly so.

× ½
Lapsana communis

× ½
Arnoseris minima

## 1. Cichorium intybus L.   Chicory. Blue Sailor

Native to Eurasia and well-established as a common weed of roadsides and waste places, probably in every Ohio county. The root is often used in flavoring coffee, or sometimes as a coffee substitute. July–Oct.

Three corolla color forms can often be detected in living plants, although seldom from herbarium specimens: blue in forma *intybus*, pink to rose in forma *roseum* Neum., and white in forma *album* Neum.

x ½
Cichorium intybus

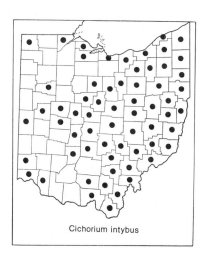

Cichorium intybus

Krigia virginica

Krigia dandelion

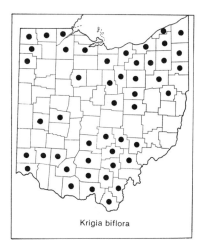

Krigia biflora

## 65. Krigia Schreb. DWARF DANDELION

Heads ligulate, usually solitary. Phyllaries few to many, mostly biseriate. Receptacle naked. Flowers perfect, corollas yellow or orange. Pappus double, the outer of chaffy scales, the inner of capillary bristles. Achenes terete or somewhat angled, nerved, or ribbed, faintly transversely rugulose. Fibrous-rooted annuals or perennials with alternate, mostly basal, entire to pinnatifid leaves and milky juice. Stems either branched from the base and leafy, or unbranched and leafless. For the basic taxonomic treatment of the genus, see Shinners (1947).

a. Plants scapose; flowering stem unbranched.
  b.  Pappus bristles about 5, alternating with as many scales; plants annual
    . . . . . . . . . . . . . . . . . . . . . . . . . . . . . . . . . . . . . . . . . . . . . . .1. *K. virginica*
  bb. Pappus bristles 15–40; plants perennial. . . . . . . . . . . . . .2. *K. dandelion*
aa. Plants branched (at least the flowering stem), perennial; leaves oblanceolate, mostly entire, toothed, irregularly lobed or pinnatifid . . . . . . . . . . .
  . . . . . . . . . . . . . . . . . . . . . . . . . . . . . . . . . . . . . . . . . . . . . . .3. *K. biflora*

1. **Krigia virginica** (L.) Willd.   VIRGINIA DWARF DANDELION
Dainty annual; scapes usually several, glabrous or glandular near head; leaves basal or nearly so, linear, obovate to oblanceolate, entire to pinnatifid, sparsely hirsute to glabrous; phyllaries lanceolate, reflexed with age; achenes many-nerved or -ribbed; pappus of 5–7 scales alternating with as many (or twice as many) scabrous bristles. Common in sterile soils in a few counties. March–July.

2. **Krigia dandelion** (L.) Nutt.   POTATO-DANDELION
  *Cynthia dandelion* (L.) DC.
Perennial, scapose, tuber-bearing, short, barely 12 cm tall; leaves basal, lanceolate to oblanceolate, toothed to pinnatifid, glabrous to scarcely pubescent; phyllaries 5–10, mostly glabrous or with a few glandular hairs, becoming reflexed with age; pappus of 15–40 upwardly barbed capillary bristles, scales obscure. Dry habitats, chiefly in sandy soils. Reported only from Adams County in southern Ohio, where it was found in a thin, oak woods. April–June.

3. **Krigia biflora** (Walt.) Blake
  *Cynthia virginica* (L.) Don
  *Krigia amplexicaulis* Nutt.
Perennial; scapes branched and bearing 1–several heads and a few reduced, small, clasping leaves; leaves mostly basal, oblanceolate, entire or toothed to pinnatifid, glabrous; phyllaries 10–20, acuminate, shorter than the orange corollas, reflexed with age; pappus of 20–30 or more fragile bristles and a few scales. Common in open woods and fields. April–Aug.

## 66. Hypochaeris L. CAT'S-EAR

Heads showy, terminating rather stout scapes. Phyllaries in several series, often hispid along dorsal crest, the outer smaller, becoming reflexed with age. Receptacle chaffy. Flowers ligulate, perfect, yellow. Pappus of plumose bristles, the outer often merely barbellate. Achenes with a long slender beak. Perennial herbs; scapes usually branched, glabrous above, hirsute or spreading-hispid

×1

×5

× ½

Krigia virginica

× ¼

Krigia dandelion

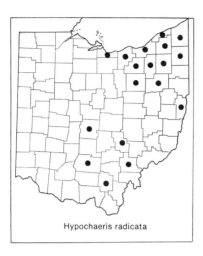

Hypochaeris radicata

below. Leaves hispid, mostly basal, or the stems somewhat leafy below; leaf margins entire, toothed, or pinnatifid.

1. HYPOCHAERIS RADICATA L.    LONG-ROOTED CAT'S-EAR

Often a weed in fields, waste places, lawns, and other grassy places. Widely established native of Eurasia. May–Sept.

× 3

× ½ S.A.
Krigia biflora

× 2

× ½
Hypochaeris radicata

## 67. Leontodon L. FALSE DANDELION. HAWKBIT

Heads of ligulate, perfect, yellow flowers. Pappus entirely of plumose bristles, or the outer mixed with straight bristles and scales. Perennials with ovate-lanceolate to oblanceolate, mostly pinnatifid or deeply lobed leaves. Scapes branched or unbranched, leafless or with small bracts. A cosmopolitan genus of about 100 species, found especially in South America. It is not well-represented in the Ohio flora, perhaps due to lack of collecting. In addition to the species below, I have seen two specimens of *L. leysseri* (Wallr.) G. Beck at the Kent State University Herbarium. This species is introduced from Europe; its existence in the Ohio flora today is doubtful and it is not included in the key.

a. Scapes branched, and usually with small bracts; pappus of plumose bristles . . . . . . . . . . . . . . . . . . . . . . . . . . . . . . . . . . . . . . . .1. *L. autumnalis*

aa. Scapes unbranched, usually without bracts; pappus on marginal achenes of scales mixed with barbellate hairs . . . . . . . . . . . . . . . . .2. *L. hispidus*

× 1
Leontodon autumnalis

× ½
Leontodon hispidus

Leontodon autumnalis

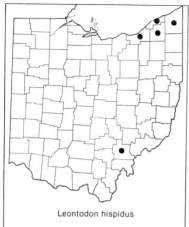

Leontodon hispidus

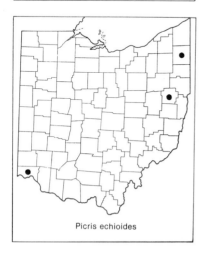

Picris echioides

### 1. Leontodon autumnalis L.   Fall Dandelion. Hawkbit

Perennial from stout caudex; stems scapose, 30–40 cm tall; scape branched, glabrous or puberulent below heads; leaves basal, oblanceolate, glabrous to sparsely hirsute, lobed, usually deeply so; heads solitary; phyllaries narrow, imbricate; achenes transversely rugulose, fusiform; pappus of plumose bristles. Introduced from Eurasia; in Ohio a somewhat rare plant of fields, lawns, waste places, and roadsides. Its rarity in herbarium collections is perhaps due to its similarity to the common dandelion. June–July.

The variety *pratensis* (Less.) Koch cannot be separated from the typical variety *autumnalis*, at least in Ohio plants.

### 2. Leontodon hispidus L.   Common Hawkbit

Perennial to 40 cm tall; scapose, the scapes simple and unbranched, mostly hispid but occasionally nearly glabrous; leaves oblanceolate, pinnately lobed, deeply or shallowly toothed; phyllaries imbricate; achenes scabrous, fusiform; pappus of 2 kinds, the inner plumose, the outer of slender scales or barbellate bristles. Introduced from Europe, this species occurs in much the same type of habitats as *L. autumnalis*, but is the rarer of the two. Summer–fall.

### 68. Picris L.   BITTERWEED

Flowers ligulate, perfect, corollas yellow. Heads many-flowered, terminating leafy stems. Phyllaries in 2 or more series; achenes incurved, subterete, usually ribbed, often obscurely so, beaked or beakless, rugulose. Pappus of 1 or 2 rows of plumose bristles. Coarse, rough, bristly annuals or perennials with alternate, entire to somewhat pinnatifid leaves. A genus of about 50 species, all native to Eurasia and the Mediterranean regions. Both species that occur in Ohio have been found but rarely.

a. Outer phyllaries narrow; achenes beakless; pappus sparsely plumose . . . . .
. . . . . . . . . . . . . . . . . . . . . . . . . . . . . . . . . . . . . . . . . . . . . . . .1. *P. hieracioides*
aa. Outer phyllaries ovate or ovate-lanceolate; achenes decidedly beaked; pappus densely plumose . . . . . . . . . . . . . . . . . . . . . . . . . . . . . . . .2. *P. echioides*

### 1. Picris hieracioides L.

Perennial, 0.5 to 1 m tall; stems spreading-hispid; cauline leaves lanceolate, sessile, irregularly toothed, barely clasping, reduced upward; heads in a corymbose arrangement; phyllaries imbricate, narrow, less than 3 mm wide; achenes obscurely rugulose, rather abruptly narrowed but beakless; pappus sparsely plumose. Native to Eurasia, known in Ohio from only a few collections. It grows along railroad tracks and roadsides and in waste places. July–Aug.

### 2. Picris echioides L.   Ox-tongue. Bristly Picris

Annual; stems coarse, spiny-hispid, to 0.5 m tall; leaves spinescent, toothed to entire, lanceolate, sessile, sub-clasping; outer phyllaries ovate, somewhat folia-

x 1

x 3

x 1

x 2

x ½
Picris echioides

x ½
Picris hieracioides

Picris hieracioides

ceous, spinescent, the inner about the same length, narrower than the outer, thickened below; achenes decidedly beaked, curved, partly enclosed by the phyllaries, rugulose. Native to Europe, known in Ohio from a few collections from open fields. Its existence in Ohio today is doubtful.

### 69. Tragopogon L.  GOAT'S-BEARD

Heads of ligulate, perfect, yellow, pale yellow, or purple flowers. Heads solitary. Phyllaries in a single series and equal. Receptacle naked. Achenes terete or slightly angled, beaked. Pappus a single series of plumose bristles. Perennial or biennial taprooted herbs. Leaves alternate, entire, glabrous, linear or lanceolate, clasping at base.

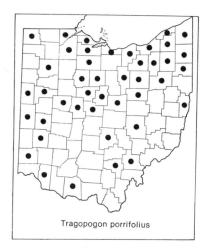

Tragopogon porrifolius

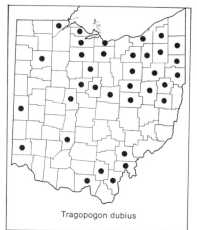

Tragopogon dubius

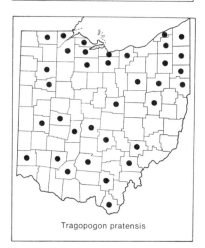

Tragopogon pratensis

A genus of approximately 50 species, native to Eurasia and northern Africa. The species occasionally hybridize, and as a result of chromosome doubling in the diploid hybrids, establish true-breeding tetraploid races. These were first reported by Ownbey (1950) and later verified chemically by Brehm and Ownbey (1965).

Characteristically, the flowers close at noon, hence the name "John-go-to-bed-at-noon" applied by the English.

a. Flowers purple, equalling or shorter than the phyllaries . . . . . . . . . . . . . .
. . . . . . . . . . . . . . . . . . . . . . . . . . . . . . . . . . . . . . . . . . . . . . .1. *T. porrifolius*

aa. Flowers yellow.

b. Peduncles strongly inflated below heads; corollas shorter than the phylla
ries; leaves tapering uniformly from base . . . . . . . . . . . . . . .2. *T. dubius*

bb. Peduncles scarcely or not at all inflated below heads; corollas equalling
or longer than the phyllaries; leaves abruptly narrowed at base, their
acuminate tips often curled downward at maturity . . . . . .3. *T. pratensis*

1. Tragopogon porrifolius L.   Salsify. Vegetable Oyster

Biennial, stems to 1 m tall; leaves glabrous, gradually tapering from base to apex; peduncles enlarged below the head, especially in fruit; phyllaries equalling or exceeding the purple corollas; achenes long and abruptly contracted to a long beak; pappus tawny. Introduced from Europe and widely established in Ohio, where it occurs as a weed in fields and meadows, and along roadsides. Cultivated in Europe and occasionally in the United States for its edible taproot. June–Aug.

2. Tragopogon dubius Scop.

*T. major* Jacq.

Biennial; stems to 1 m tall; peduncles inflated below head; leaves glabrous, gradually tapering from the slightly enlarged base, sparsely pubescent in axils; phyllaries longer than the distinctively pale yellow corollas; achenes tapering to beak at apex; pappus more white than tawny. Introduced from Europe and now widely established in Ohio, where it occurs in meadows and fields and along roadsides. June–Aug.

3. Tragopogon pratensis L.   Yellow Goat's-beard

Biennial or perennial; stems to 1 m or slightly taller; leaves glabrous, abruptly narrowed at base, often recurved at tips at maturity; peduncles not inflated, rather uniformly tapering; achenes slightly more contracted to a stout beak than in *T. dubius*; pappus white. Introduced from Europe and widely established in meadows, fields, and waste places. June–Aug.

× 1

× ½
Tragopogon dubius

× ½
Tragopogon porrifolius

× ½
Tragopogon pratensis

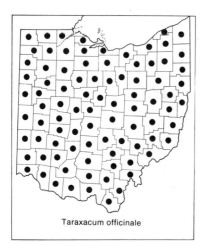

Taraxacum officinale

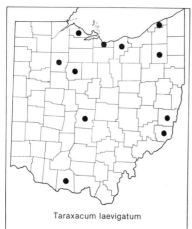

Taraxacum laevigatum

## 70. Taraxacum Wiggers  DANDELION

Corollas ligulate, perfect, yellow; heads 1–several, solitary, peduncles glabrous, hollow. Phyllaries in 2 rows, the outer shorter, the inner reflexed at maturity. Receptacle naked. Achenes fusiform, terete or slightly angled; longitudinally ribbed, muricate or tuberculate above, tannish-brown to reddish or reddish-purple. Pappus of white capillary bristles at apex of slender beak. Perennial, scapose, tap-rooted herbs with all basal, entire to pinnatifid leaves.

Native to the northern hemisphere and South America. Well over a thousand species have been described but this excessive number is probably due to apomixis, polyploidy, and hybridization which are prevalent in the genus. A detailed study might reduce this number significantly.

The thick perennial root is crowned by a short stem from which the basal leaves develop. At the beginning of growth in the spring, a terminal or near-terminal bud develops and begins active growth for that year. During eradication efforts the thick tap root must be severed very low in the ground, or otherwise a lateral bud merely gives rise to another plant. The plants are bothersome weeds in lawns. The dispersal mechanism of the fruits and their dissemination by wind enable species of this genus to reach heights of reproductive efficiency seldom attained elsewhere in the plant kingdom.

a. Achenes tan to brown or olivaceous at maturity . . . . . . . . .1. *T. officinale*
aa. Achenes reddish or reddish-purple at maturity . . . . . . . .2. *T. laevigatum*

### 1. TARAXACUM OFFICINALE Weber  COMMON DANDELION

Same characters as genus, but with achenes tan, brown, or olivaceous in color. Probably the most common species of vascular plant in Ohio, occurring in every county. Naturalized from Eurasia and found in many disturbed habitats. March–Dec.

### 2. TARAXACUM LAEVIGATUM (Willd.) DC.  RED-SEEDED DANDELION
   *T. erythrospermum* Andrz.

Similar to *T. officinale*, but differing in color of achenes, which are reddish or purplish at maturity. In addition to its distinguishing fruit color, it is of somewhat smaller stature than *T. officinale*, and has more deeply pinnatifid leaves. Less common than *T. officinale*, it also is naturalized from Eurasia and grows in various disturbed, often shaded, habitats (not illustrated). April–July.

The common name, "Red-seeded Dandelion," is an error in choice of terms. The colored fruit is a single-seeded achene and therefore, a better name would be "Red-fruited Dandelion."

## 71. Sonchus L.  SOW-THISTLE

Flowers ligulate, perfect, yellow. Heads usually many, in a corymbose or umbellate arrangement, ovoid to campanulate, many-flowered, often in excess of 150 flowers per head. Phyllaries multiseriate. Achenes glabrous, obcompressed, few- to many-ribbed, often transversely rugose, tapering at the apex, beakless. Pappus of many soft, white, capillary bristles, usually falling as a unit. Coarse, annual or perennial herbs with milky juice. Leaves alternate or basal, entire to pinnatifid, usually auriculate, prickly-margined.

As many as 50 species from Eurasia, the Mediterranean, and tropical Africa, and now naturalized in the United States. Quite often very troublesome weeds in farming regions.

a.  Plants perennial; flowering heads 2–3 cm wide; achenes 2–3 mm long; corollas bright yellow or orange.

b.  Phyllaries and peduncles with glandular hairs; involucre broadly campanulate or hemispheric ...........................1. *S. arvensis*

x ½

Taraxacum officinale

× ½
Sonchus oleraceus

× ½
Sonchus arvensis

× ½
Sonchus uliginosus

bb. Phyllaries and peduncles essentially glabrous or at most scantly tomentose; involucre slender-turbinate to cylindric .........2. *S. uliginosus*

aa. Plants annual; flowering heads small, 1–2 cm wide; achenes 1–2 mm long;
corollas pale yellow.

  b.  Leaf auricles pointed; achenes transversely rugose and with several
ribs .........................................3. *S. oleraceus*

  bb. Leaf auricles rounded; achenes 3–ribbed on each side, otherwise smooth
...............................................4. *S. asper*

## 1. SONCHUS ARVENSIS L.  FIELD SOW-THISTLE

Perennial; stems to 1.5 m tall, underground rootstalks extensive; leaves
runcinate-pinnatifid, margins subentire to spiny-toothed, bases cordate and
clasping; heads few to many, peduncles long, glandular-hispid; involucre
broadly campanulate to hemispheric; phyllaries imbricate in about three series, glandular, usually lead-colored; achenes transversely wrinkled on ribs.
Native to Europe and widely naturalized in the United States, where it occurs
along roadsides, on lake shores, and in waste places. Early summer–late fall.

## 2. SONCHUS ULIGINOSUS Bieb.

Perennial; similar to *S. arvensis*, with which it should possibly be considered
conspecific, but glabrous and a paler green; heads more slender, turbinate to
cylindric; phyllaries pale, glabrous. A very common weed naturalized from
Europe. Early summer–fall.

## 3. SONCHUS OLERACEUS L.  COMMON SOW-THISTLE

Annual; stems glabrous, to 1.5 m tall; leaves runcinate-pinnatifid or undivided
and toothed, margins prickly, auriculate, the auricles sharply acute or pointed,
cauline leaves reduced upward; heads small, rarely 2 cm wide and usually less
than 1.5 cm, heads in a corymbiform arrangement; achenes transversely rugose and also ribbed on each face. Naturalized from Europe; common in fields
and waste places. Early summer–fall.

## 4. SONCHUS ASPER (L.) Hill  SPINY-LEAVED SOW-THISTLE

Similar to *S. oleraceus*; stems glabrous or glandular with red-tipped hairs;
leaves less pinnatifid, often completely undivided, strongly spiny-toothed, auricles of the clasping base rounded; achenes 3-ribbed on each side, otherwise
smooth. Naturalized from Europe; common along roadsides, in fields and
waste places. Early summer–October.

    Three forms can be separated in our range.

a.  Cauline leaves pinnatifid.

  b.  Upper stems, peduncles, and phyllaries glandless.........forma *asper*

  bb. Upper stems, peduncles, and phyllaries stipitate-glandular............
.......................................forma *glandulosus* Beckh.

aa. Cauline leaves undivided ...............forma *inermis* (Bisch.) Beck

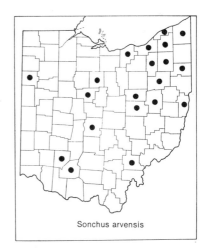

Sonchus arvensis

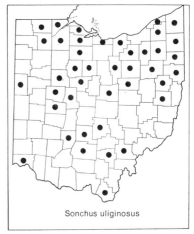

Sonchus uliginosus

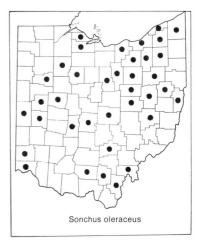

Sonchus oleraceus

Tribe X. Lactuceae (Cichorieae)  239

×½

Sonchus asper

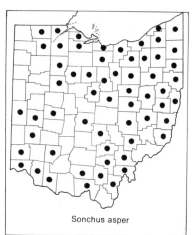

Sonchus asper

## 72. Lactuca L. LETTUCE

Flowers ligulate, perfect, yellow, cream, blue, or white. Involucre cylindric, phyllaries in two or more series of unequal lengths. Receptacle naked. Achenes flattened or at least compressed, few- to many-nerved, sometimes beaked. Pappus of capillary bristles, white to tawny brown. Annual or biennial plants with lactiferous ducts. Leaves alternate, entire to pinnatifid. A genus of as many as 100 species, chiefly in temperate Eurasia, and extending into South Africa.

In addition to the six species below, *L. sativa* L., GARDEN LETTUCE, can occasionally be found as an escape. It is easily distinguished by its broad, toothed leaves. Leaves of the other species are often collected in early spring and used as salad greens. I have also seen a specimen of *L. pulchella* (Pursh) DC. in the Kent State University Herbarium collected on railroad ballast in Trumbull County by A. N. Rood, but the species is not included here since its existence in the wild today is doubtful.

a. Corollas yellow.
  b. Achenes with 1–3 ribs on each face; flowers usually more than 12 per head.
    c. Leaves and stem glabrous or only sparsely pubescent, often glaucous; heads small, 10–15 mm high; achenes 4.5–6.5 mm long; widespread in Ohio . . . . . . . . . . . . . . . . . . . . . . . . . . . . . . . . . . . . . . . . . . .1. *L. canadensis*
    cc. Leaves and stem pubescent; heads 15–22 mm high; achenes 7–10 mm long; rare in Ohio . . . . . . . . . . . . . . . . . . . . . . . . . . . . .2. *L. hirsuta*
  bb. Achenes with 5–7 ribs on each face; flowers often fewer than 12 per head.
    c. Midribs of leaves conspicuously spinulose, their margins spinulose-toothed . . . . . . . . . . . . . . . . . . . . . . . . . . . . . . . . . . . . . . . . . .3. *L. serriola*
    cc. Midribs of leaves usually white and not spinulose, their margins not conspicuously spinulose-toothed . . . . . . . . . . . . . . . . . . . . .4. *L. saligna*
aa. Corollas blue or cream (occasionally yellowish in *L. biennis*); achenes tapering to a stout or firm beak, much shorter than the body of the achene, or achenes beakless.
  b. Pappus white; inflorescence open . . . . . . . . . . . . . . . . . . .5. *L. floridana*
  bb. Pappus brownish or tan; inflorescence narrowly paniculiform . . . . . . . . . . . . . . . . . . . . . . . . . . . . . . . . . . . . . . . . . . . . . . . . . . . . . . . . . . . . . .6. *L. biennis*

1. **Lactuca canadensis** L.   WILD LETTUCE
  *L. sagittifolia* Ell.
  *L. steelei* Britt.

Biennial to 3 m tall, stems usually glabrous; leaves entire, toothed, or pinnatifid, sagittate or abruptly tapering at base; heads many, small, 10–15 mm high, 12–20-flowered; corollas yellow; achenes very dark, rugulose, beak usually shorter or no longer than body. Common in fields, waste places, and fence rows. Late June–Sept.

Four weakly separated varieties can be recognized, but they seem not to have any distributional significance.

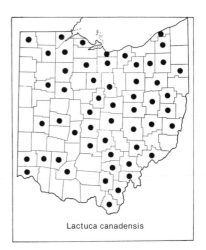

Lactuca canadensis

a. Leaves mostly unlobed, entire or toothed.

b. Cauline leaves lanceolate or lance-ovate, entire or infrequently toothed
..............................................var. *canadensis*

bb. Cauline leaves oblanceolate or obovate, nearly always toothed........
..........................................var. *obovata* Wieg.

aa. Leaves, or most of them, lobed to pinnatifid.

× 2

× 2

× 3

× ½

× ½

Lactuca canadensis
var. canadensis

Lactuca canadensis
var. latifolia

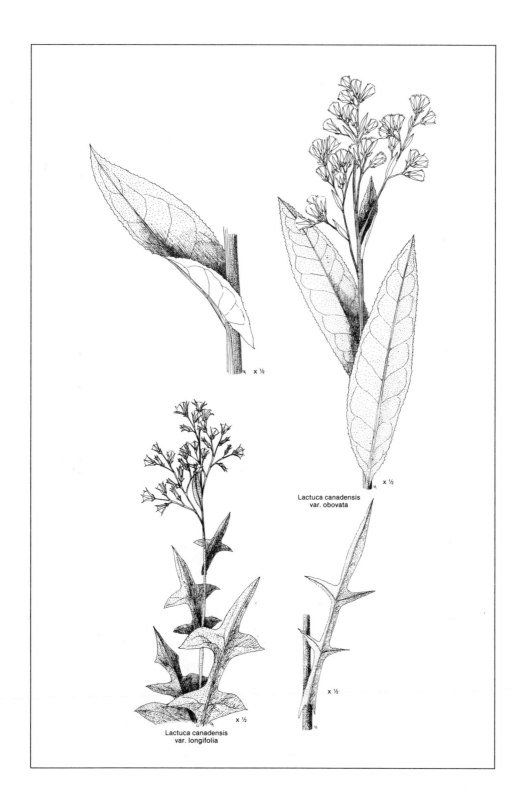

Lactuca canadensis
var. obovata

Lactuca canadensis
var. longifolia

b.  Lobes linear-falcate, entire or usually so; the unlobed upper leaves, if any, linear to linear-lanceolate .............var. *longifolia* (Michx.) Farw.

bb.  Lobes broadly falcate to obovate, entire or toothed; the unlobed upper leaves, if any, lanceolate or lance-ovate ..........var. *latifolia* Kuntze

2.  **Lactuca hirsuta** Muhl.   Hairy Lettuce

Biennial to 3.5 m tall; stems sparsely villous; leaves as in *L. canadensis* except copiously pubescent on midribs beneath and sparsely so on surface. Rare in Ohio, where it occurs in fields, usually near woodlands. July–Sept.

3.  Lactuca serriola L.   Prickly Lettuce

   *L. scariola* L.

Biennial to 1.5 m tall; stems prickly below, glabrous above; leaves prickly on midrib beneath and on margins, lobeless or pinnately lobed, sagittate-clasping, the basal leaves often twisted and oriented to sun; heads numerous, rarely over 12-flowered, corollas yellow (sometimes blue when dry); achenes with 5–7 nerves on each face, spinulose near summit, beak about the same length as body. Naturalized from Europe, a common weed in fields and waste places. July–Sept.

Two varieties can be recognized among Ohio plants.

a.  Leaves pinnately lobed ...............................var. *serriola*

aa.  Leaves unlobed .....................var. *integrata* Gren. & Godr.

4.  Lactuca saligna L.   Willow-leaved Lettuce

Biennial or annual, stem to 0.8 m tall, glabrous; leaves sagittate, linear and entire or pinnatifid, glabrous, midvein usually white; heads many, arranged in a usually slender, elongate panicle, usually 8–12-flowered, corollas yellow; achenes with 5–7 nerves on each face, scabrous near summit, beak twice as long as body. A weedy species widely distributed in Ohio. Introduced and naturalized from Europe. July–Sept.

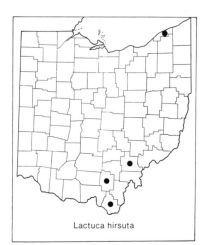

Lactuca hirsuta

Lactuca serriola

×½
Lactuca hirsuta

×2

×½
Lactuca serriola

Tribe X. Lactuceae (Cichorieae)   245

### 5. **Lactuca floridana** (L.) Gaertn.   Tall Blue Lettuce. Woodland Lettuce

Annual or more often biennial; stems to 2 m tall; leaves elliptic to cordate, glabrous or pubescent along main veins, the margins toothed to lobed; heads many, arranged in a large spreading panicle, 10–20-flowered, corollas blue; achenes beakless or with a short, stout and curved beak, and with 5–7 nerves

× ½

Lactuca saligna

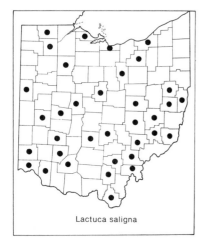

Lactuca saligna

on each face; pappus white. Common in moist ditches and swales, and at woodland borders. July–Sept.

Two varieties can be recognized among Ohio plants: var. *floridana* with pinnatifid leaves, and var. *villosa* (Jacq.) Cronq. with elliptic to cordate leaf blades and leaf margins that are merely toothed.

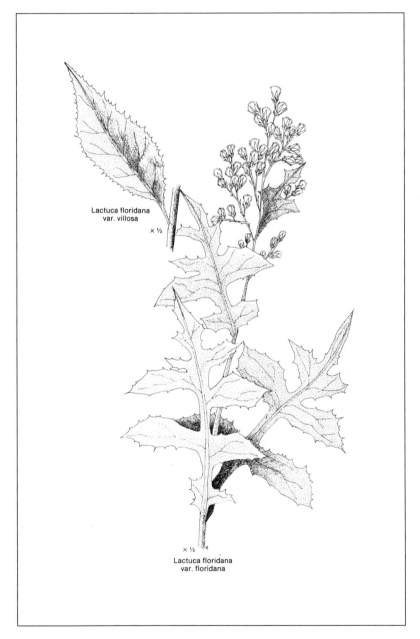

Lactuca floridana
var. villosa

× ½

× ½

Lactuca floridana
var. floridana

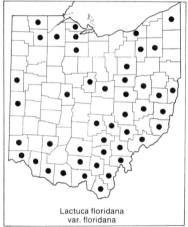

Lactuca floridana
var. floridana

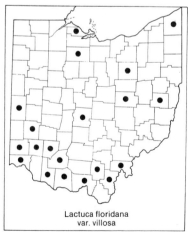

Lactuca floridana
var. villosa

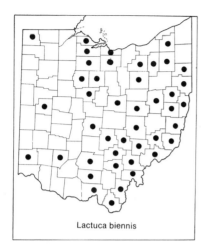

Lactuca biennis

**6. Lactuca biennis** (Moench) Fern.   Tall Blue Lettuce

Large annual, or more often biennial; stems to 3.5 m tall; leaves pinnatifid or merely toothed, sometimes sagittate, glabrous or pubescent along the main veins beneath; heads many, arranged in an elongated panicle; corollas bluish to white, or sometimes yellowish; achenes with several nerves on each face, essentially beakless or with a short, stout beak; pappus light brown. Rather common in moist habitats. July–Sept.

In addition to forma *biennis*, with lobed leaves and bluish flowers, two other forms can be distinguished. These are forma *integrifolia* (T. & G.) Fern. with leaves toothed but unlobed, and forma *aurea* (Jennings) Fern. with corollas yellow in color.

*Lactuca* x *morsii* Robins is the apparent hybrid of *L. biennis* x *L. canadensis*. Although both species occur in Ohio, I have not found this hybrid among the specimens I have examined.

### 73. Crepis L.   hawk's-beard

Heads in panicles or corymbs. Flowers ligulate, perfect, yellow. Phyllaries in 2 series, the outer much reduced. Achenes terete or slightly flattened, fusiform, 10–20-ribbed, slender-beaked or not. Pappus of soft, white, capillary bristles. Taprooted annuals or perennials with lactiferous ducts, our species without rhizomes. Leaves alternate, entire to pinnatifid, mostly basal and reduced upward to mere bracts. Advanced students should be familiar with the extensive cytotaxonomic work on the genus by Babcock (1947).

A few specimens of *C. biennis* L., *C. nicaeensis* Balb., and *C. tectorum* L. are deposited in The Ohio State University Herbarium and a few other herbaria, but are considered waifs and not included here. However, since all our species are introduced and somewhat weedy, these and other species can be anticipated.

a.   Stems and leaves glandular-pubescent. . . . . . . . . . . . . . . . . .1. *C. pulchra*
aa.  Stems and leaves glabrous or pubescent, not glandular. . . .2. *C. capillaris*
1. Crepis pulchra L.

Annual, 0.5–1 m tall; stems spreading-glandular below to essentially glabrous above; leaves spreading-glandular, the lower oblanceolate, coarsely toothed to pinnatifid; heads several, arranged in a somewhat corymbose manner; outer phyllaries glabrous, short, the inner longer with thickened and raised midrib; achenes slightly contracted at the apex, 10–12-ribbed; pappus soft, white, capillary. Native to Europe; found in waste places and along roadsides, mostly in the southern part of Ohio. May–June.

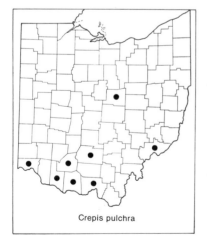

Crepis pulchra

Lactuca biennis
forma biennis

× 3

× ½

Lactuca biennis
forma integrifolia

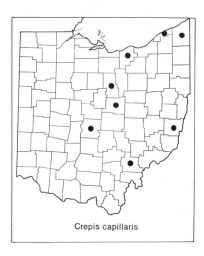

Crepis capillaris

**2. CREPIS CAPILLARIS (L.) Wallr. SMOOTH HAWK'S-BEARD**
Annual; stems usually much-branched, mostly glabrous; leaves mostly glabrous, occasionally hispidulous, basal leaves lanceolate to oblanceolate, denticulate to pinnatifid, the upper reduced and often clasping; heads several; phyllaries hairy, sometimes glandular, the inner thickened on back, longer than the outer; achenes contracted at apex, about 10-ribbed. Native to Europe; found primarily in northern and eastern counties in lawns, pastures, and waste places. Uncommon. July–Oct.

**74. Prenanthes L. WILD LETTUCE. RATTLESNAKE-ROOT**
Flowers ligulate, perfect, corollas pink, purplish, cream, greenish, or yellowish-white. Phyllaries in a single whorl with a few small ones below, the smaller often grading into the larger. Achenes short, linear-oblong, striate, glabrous, reddish-brown. Pappus of copious whitish or brownish, rough, deciduous, capillary bristles. Perennials with tuber-like roots. Leaves alternate.

About 40 species of North America, tropical Africa, and Eurasia. This treatment is based in part on that of Milstead (1964), according to whom the taxonomy is often clouded and uncertain due to frequent hybridizations. Fernald (1950) attributed *P. trifoliata* (Cass.) Fern. to Ohio, but I have seen no specimens.

a. Heads and clusters of heads borne close to the stem.
  b. Corollas pink to purplish; stem and leaves glabrous and glaucous (main rachis of inflorescence often soft-pubescent) ...........1. *P. racemosa*
  bb. Corollas creamy; upper stems and leaves rough-hairy or scabrous ......
    ...............................................2. *P. aspera*
aa. Heads and clusters of heads divergent from the main stem.
  b. Phyllaries pubescent with at least a few long, scattered, coarse hairs.
    c. Primary phyllaries about 8; flowers fewer than 15 per head ..........
    ........................................3. *P. serpentaria*
    cc. Primary phyllaries 12–15; flowers 20 or more per head ............
    .......................................4. *P. crepidinea*
  bb. Phyllaries glabrous or essentially so.
    c. Heads small, 5–6-flowered; phyllaries usually 5; pappus whitish to yellow-brown....................................5. *P. altissima*
    cc. Heads 8–15-flowered; phyllaries usually 7–10; pappus reddish-brown
    ...............................................6. *P. alba*

**1. Prenanthes racemosa Michx.**
Stems to 1.5 m tall, smooth, glabrous, glaucous, often pubescent in branches of the inflorescence; lower leaves petiolate, oblanceolate, the cauline and upper reduced, becoming lanceolate, toothed, sessile, and clasping; heads ar-

Prenanthes racemosa

× 3

× ½
Crepis pulchra

× ½
Crepis capillaris

Prenanthes aspera

ranged in elongate clusters forming a dense or interrupted thyrse; heads about 10–20-flowered, corollas pink to purplish; phyllaries 7–14, dark at maturity, scantly or densely long-hairy; pappus yellowish-tan. Calcareous bogs, wet prairies and meadows. Aug.–Sept.

2. **Prenanthes aspera** Michx.

Very similar to *P. racemosa*, above, except for its rough-pubescent stems and leaves, usually more dense inflorescences, larger heads, and cream-colored flowers. Dry prairies and similar dry places. Rare in Ohio. Aug.–Sept.

3. **Prenanthes serpentaria** Pursh   LION'S-FOOT

Stems to 1 m tall, glabrous, often purplish; leaves glabrous, pinnately lobed (occasionally unlobed), the lobes rounded; heads clustered at the ends of elongated branches, nodding; corollas yellowish, creamy, or pinkish; phyllaries

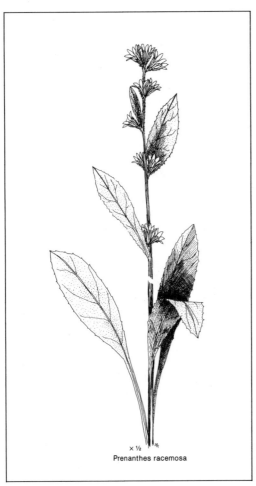

× ½
Prenanthes racemosa

× ½
Prenanthes aspera

with at least a few long hairs, shorter than the creamy pappus. Dry areas in southern Ohio. Aug.–Sept.

x 1

× ½    3A

Prenanthes serpentaria

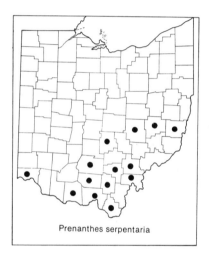

Prenanthes serpentaria

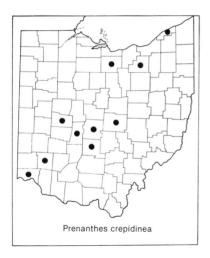

Prenanthes crepidinea

Prenanthes altissima

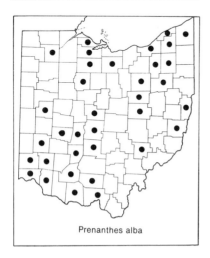

Prenanthes alba

**4. Prenanthes crepidinea** Michx.   Corymbed Rattlesnake-root

Stems to 2 m or more tall, smooth, glabrous to pubescent on branches of inflorescence; leaves glabrous, occasionally somewhat scabrous and hirsute on midrib beneath, lower leaves large, triangular-ovate, hastate or cordate to shallowly lobed, upper leaves gradually reduced, ovate, and irregularly toothed; inflorescence corymbose to paniculiform; heads usually large, nodding, 25–35-flowered, corollas creamy; phyllaries with at least a few hairs, the outer reduced and few; pappus light tan to brown. Moist woodland openings and borders. Aug.–Sept.

**5. Prenanthes altissima** L.

Stems 0.5 to 2 m tall, glabrous, occasionally pubescent near base; lower leaves long-petiolate, leaves reduced upward, quite variable in shape and size, somewhat membranous, ovate, cordate, triangular or frequently 3–5-parted or -lobed; inflorescence elongate-paniculiform to more open and loose; heads small, nodding, 5–6-flowered, corollas greenish-white; phyllaries mostly 5, often with darkened tips; pappus tannish-white to yellowish-brown. Common in Ohio, where it occurs in moist, open woodlands. July–Oct.

**6. Prenanthes alba** L.   White Lettuce

Stems smooth, glaucous, 0.5 to 1.5 m tall; leaves glabrous above, occasionally pubescent beneath, variable in shape: triangular-hastate, sinuate toothed, or 3–5-lobed; inflorescence open-paniculiform, heads nodding, 8–5-flowered, corollas purplish to greenish; phyllaries 7–10, glabrous, pappus reddish-brown. Woodlands. Aug.–Sept.

### 75. Hieracium L.   Hawkweed

Heads in panicles or corymbs. Flowers ligulate, perfect, yellow, orange or orange-red. Phyllaries in 2–3 series, or more often in a single series, usually glandular. Achenes terete to slightly angled, not or only slightly tapering at the apex, but not beaked, mostly narrowed at the base, faintly ribbed. Pappus of a single series of capillary, often rough, gray, tawny, or brown bristles. Perennials. Leaves alternate or all basal, few to many, entire or toothed, usually pubescent and often glandular-stipitate.

Perhaps several thousand species comprise this widely distributed and variable genus, many of which are quite variable due to hybridization, apomixis, and polyploidy. The introduced species, primarily from Europe, and most of our native species, maintain their characteristics quite well. A few species are cultivated and, although attractive in groups, spread rapidly by seed and runners and very soon become pervasive weeds.

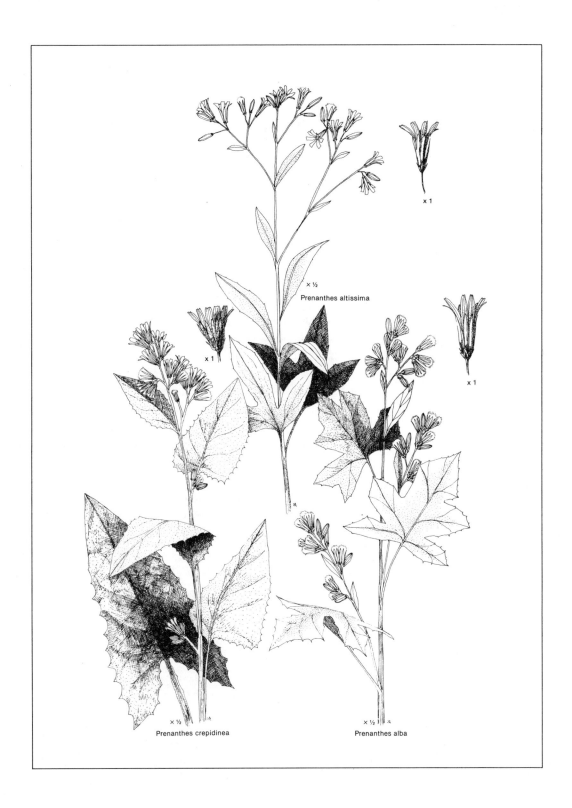

x 1

× ½
Prenanthes altissima

x 1

x 1

× ½
Prenanthes crepidinea

× ½
Prenanthes alba

Tribe X. Lactuceae (Cichorieae)   255

a. Basal rosettes present at anthesis; cauline leaves none or few.

  b. Corollas red-orange, becoming deep red when dry ...1. *H. aurantiacum*

  bb. Corollas yellow.

    c. Scapes bearing only 1–3 heads (usually 1)............2. *H. pilosella*

    cc. Scapes bearing 5 or more heads.

      d. Basal leaves narrowly oblanceolate to narrowly lanceolate, lateral veins obscure.

        e. Leaf blades glabrous or sparsely hairy above and on margins, usually somewhat glaucous.

          f. Plants stoloniferous, and rhizomes abundant and elongate; leaves broadly ovate-lanceolate.....................3. *H. floribundum*

          ff. Plants not stoloniferous, rhizomes short and thick; leaves narrowly lanceolate ...............................4. *H. florentinum*

        ee. Leaf blades abundantly pubescent on both surfaces, not glaucous .........................................5. *H. caespitosum*

      dd. Basal leaves ovate to elliptic, lateral veins evident.

        e. Blades thin, often with purple veins, somewhat glaucous, essentially glabrous.....................................6. *H. venosum*

        ee. Blades thick, green, not glaucous, pubescent on both surfaces .....
.............................................7. *H. trailii*

aa. Basal rosettes usually absent at anthesis; stems leafy.

  b. Phyllaries decidedly imbricated in at least 2 series; leaves irregularly dentate; plants found only in northern counties ........8. *H. canadense*

  bb. Phyllaries in a single series or a few very small bracts also sometimes present; leaves entire or merely denticulate.

    c. Hairs on the lower part of the stem at least 1 cm, often longer; a rare species found only in northern counties ...........9. *H. longipilum*

    cc. Hairs on the lower part of the stem 1 cm long or less.

      d. Stems slender; heads small and arranged in an open panicle with flexuous branches; leaves thin, glaucous beneath ................
.........................................10. *H. paniculatum*

      dd. Stems thicker; heads in an elongate or condensed panicle.

        e. Stems leafy chiefly at the base, sparsely pubescent, sparingly stipitate-glandular in inflorescence; achenes tapering at summit....
.........................................11. *H. gronovii*

        ee. Stems leafy throughout, pubescent, densely stipitate-glandular in inflorescence; achenes truncate ..................12. *H. scabrum*

1. HIERACIUM AURANTIACUM L. DEVIL'S PAINT BRUSH. KING DEVIL
Perennial with many basal rosettes and slender rapidly spreading stolons; stems
to 0.5 m tall, naked or with 1 or 2 small bracts, long-hairy, stellate-tomentose

× 1

Hieracium floribundum
× ½

Hieracium pilosella
× ½

Hieracium aurantiacum
× ½

and hispid with conspicuous gland-tipped hairs above; basal leaves ovate-lanceolate to oblanceolate, tips blunt or rounded, long-hairy on both surfaces; heads several to 25 in a tight corymb; flowers reddish-orange, becoming deep red when dried; phyllaries setose and hispid with black-tipped hairs. Native to Europe; widely distributed in the eastern part of Ohio, where it occurs in fields and pastures and along roadsides. For early records of this species in Ohio, see Voss and Böhlke (1978). Common. June–Sept.

2. HIERACIUM PILOSELLA L.   MOUSE-EAR

Dwarf perennial, forming dense stoloniferous mats, rarely 35 cm tall; stems leafless, glandular-puberulent and long-hairy below, with abundant black-tipped hairs above; basal leaves oblanceolate to ovate-lanceolate, stellate-pubescent above and beneath, more densely so on margins and veins; heads usually solitary, occasionally 2 or 3, long-pedunculate; phyllaries stellate to hispid with black-tipped hairs. Native to Europe, and now well-established and often a troublesome weed in pastures and fields, but not particularly abundant in Ohio. June–Sept.

3. HIERACIUM FLORIBUNDUM Wimm. & Grab.   KING DEVIL

In habit much like H. pilosella except only sparsely pubescent upward, with more flowering heads, with generally fewer gland-tipped hairs, and with leaves somewhat glaucous. Native to Europe, an aggressive weed in the west, but uncommon in Ohio, where it occasionally occurs in fields and pastures. June–Aug.

4. HIERACIUM FLORENTINUM All.   KING DEVIL

Perennial from a short, thick rhizome; stems to 0.5 m or more tall, usually naked, becoming stipitate-glandular below heads; leaves glaucous with a few long-setose hairs, linear-lanceolate to oblanceolate; heads in a corymb; phylla-

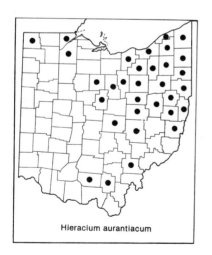

Hieracium aurantiacum

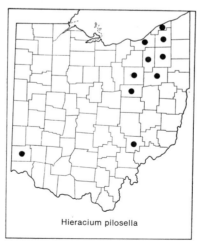

Hieracium pilosella

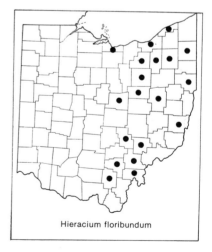

Hieracium floribundum

× ½
Hieracium venosum

× ½
Hieracium florentinum

× ½
Hieracium caespitosum

Tribe X. Lactuceae (Cichorieae)   259

ries hispid with black, gland-tipped hairs. A native of Europe with scattered distribution in eastern and northeastern Ohio, where it often occurs as an aggressive weed in fields and dry habitats. May–Sept.

5. Hieracium caespitosum Dumort.   King Devil

 *H. pratense* Tausch

Perennial from stout rhizome, occasionally with short stolons; stems to 1 m tall, long-hairy throughout and with black-tipped hairs above, naked; basal leaves oblanceolate, long-setose on both surfaces; heads several in a corymb; phyllaries hispid with black-tipped and long-setose hairs. Native to Europe, and now naturalized in our area, where it is a common weed in fields and pastures. May–Sept.

6. **Hieracium venosum** L.   Veined Hawkweed

Perennial from short rhizome; stems barely 0.5 m tall, naked or with a few reduced bract-like leaves; basal leaves elliptic, obovate, or oblanceolate, setose on margins, otherwise glabrous, veins and midribs most often red-purple; heads several in an open corymb, peduncles slender; phyllaries glabrous to stipitate-glandular. Open woods and woodland borders. May–Sept.

 A variable species which apparently hybridizes with other species, namely *H. gronovii, H. scabrum* and *H. paniculatum,* all diploids with chromosome numbers of n = 9. A few of these putative hybrids have been identified in various Ohio herbaria.

7. **Hieracium trailii** Greene

 *H. greenii* Porter & Britt.

Perennial from short thickened rhizome; stems to 0.7 m tall, densely long-setose below to sparsely so above; leaves elliptic to ovate, densely long-setose; heads many in an elongated panicle; peduncles stouter than in *H. venosum,*

Hieracium florentinum

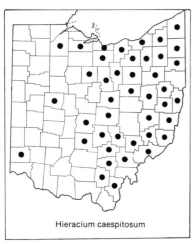

Hieracium caespitosum

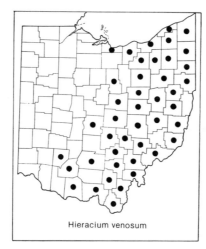

Hieracium venosum

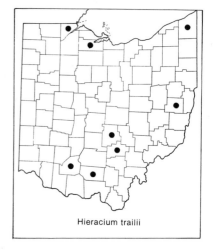

Hieracium trailii

× 1

× ½
**Hieracium longipilum**

× ½
**Hieracium trailii**

× ½
**Hieracium canadense**

1

Hieracium canadense

Hieracium longipilum

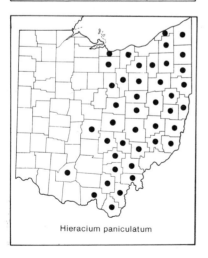

Hieracium paniculatum

densely stellate-tomentose and densely covered with black-tipped glandular hairs. Rare in Ohio, in dry open places. May–Aug.

This species is very similar to *H. venosum* but can be distinguished by its thicker and more pubescent leaves, stouter peduncles, and slightly larger heads, and by the absence of the reddish veins.

### 8. **Hieracium canadense** Michx.

Perennial; stems to 1.5 m tall, spreading-hairy or merely puberulent below to puberulent above; leaves sessile, ovate to ovate-lanceolate, irregularly toothed, stellate-puberulent or essentially glabrous above, short-hairy to puberulent beneath, the lower often nearly deciduous, the upper slightly clasping and only gradually reduced; heads many-flowered, in a corymb or more often in an umbellate arrangement, peduncles stout, puberulent; phyllaries imbricate in several series, essentially glabrous. Rare in Ohio, and apparently restricted to the northern counties, where it grows in sandy, dry soils. Aug.–Oct.

Of the three described varieties, Ohio plants belong to var. *fasciculatum* (Pursh) Fern.

### 9. **Hieracium longipilum** Torr.

Perennial from a short, thickened caudex; stems to 0.5 m tall; stems densely long-hairy below and often to the summit, hairs exceeding 1 cm; leaves erect, often crowded at base, oblanceolate to spatulate, long-hairy like stem; heads 40–90-flowered; phyllaries stellate, hispid with black, gland-tipped hairs. Dry prairies and sandy habitats. July–Aug.

This species is included here with some reluctance on the basis of a single collection by Dr. William Easterly from the Oak Openings in Lucas County in 1979, representing a new state record. However, this species is not uncommon in northern Indiana, and according to Dr. Edward G. Voss (personal communication), it is not uncommon in similar habitats in south-central and southern Michigan. It is probable that additional collections will be made from northwestern Ohio in the future.

### 10. **Hieracium paniculatum** L.   Panicled Hawkweed

Perennial; stems to 1.5 m tall, glabrous or sometimes sparsely pubescent below; leaves thin, glabrous, the lower surface glaucous, margins irregularly toothed, the lower petiolate but deciduous by anthesis, the upper gradually reduced, sessile, ovate-lanceolate; heads in open, lax, flexuous panicles, peduncles stipitate-glandular, 20–40-flowered; phyllaries glabrous or stipitate-glandular; achenes slightly tapering at apex. Open sandy woodlands. Common in eastern half of the state. July–Oct.

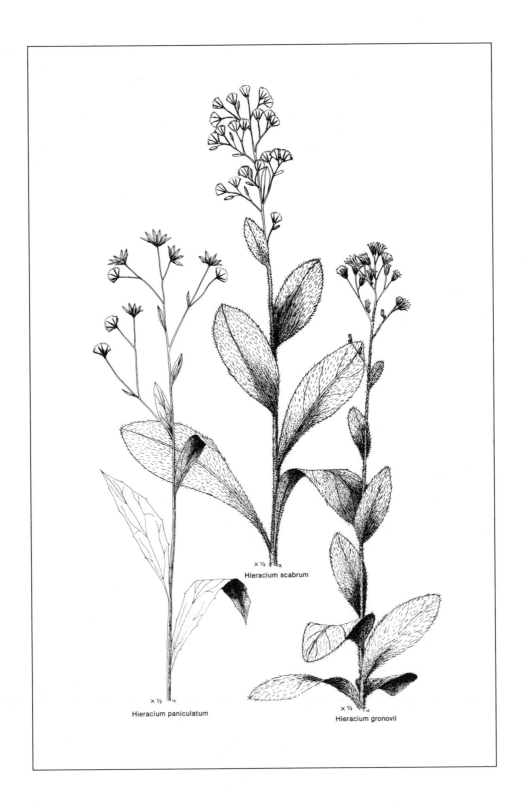

× ½

Hieracium scabrum

× ½

Hieracium paniculatum

× ½

Hieracium gronovii

## 11. **Hieracium gronovii** L.

Stems spreading-hairy toward the base, glabrous above, hairs less than 1 cm long; leaves finely stellate with a few scattered hairs, basal leaves oblanceolate to ovate, often deciduous, cauline leaves gradually reduced upwards, the uppermost sessile; heads arranged in an elongated panicle, peduncles puberulent to densely stipitate-glandular; phyllaries glabrous to stipitate-glandular; achenes tapering at summit; pappus tawny. Open woodlands, fields, and pastures, especially in sandy soils. July–Sept.

## 12. **Hieracium scabrum** Michx.   ROUGH HAWKWEED

Stems usually solitary from a short caudex, stems long-hairy below, stellate and becoming long-stipitate-glandular especially in the inflorescence; leaves thin, setose on both surfaces, densely so on petioles, margins essentially entire, leaves soon deciduous, at least the lower; heads in an open corymb or somewhat elongated panicle; phyllaries usually with blackish, gland-tipped hairs; achenes truncate, only slightly narrowed upward. Dry woodlands and woodland borders, especially in sandy and well-drained soils. July–Oct.

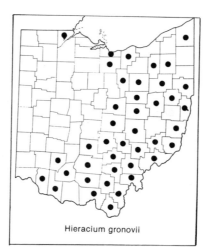

Hieracium gronovii

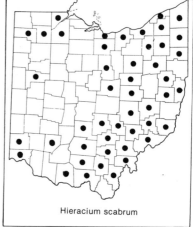

Hieracium scabrum

# LITERATURE CITED

ANDERSON, D. M. 1968. Monograph of the genus *Silphium*. I. *S. laciniatum* L. and *S. albiflorum* Gray. M.S. thesis, The Ohio State University, Columbus.

AUFFENORDE, T. M., and W. A. WISTENDAHL. 1983. Demography of *Silphium laciniatum* at the O. E. Anderson Compass-plant Prairie in southeastern Ohio, *in* R. Brewer, ed., Proc. 8th North American Prairie Conference, pp. 30–32. Dept. of Biology, Western Michigan University, Kalamazoo.

BABCOCK, E. B. 1947. The genus *Crepis*. Part I. The taxonomy, phylogeny, distribution, and evolution of *Crepis*. Univ. Calif. Publ. Bot. **21:** 1–198. Part II. Systematic treatment. ibid. **22:**199–1030.

BARKLEY, T. M. 1962. A revision of *Senecio aureus* Linn. and allied species. Trans. Kansas Acad. Sci. **65:** 318–408.

———. 1978. *Senecio, in* North American Flora, Series II, Part 10, pp. 50–139.

BARKLEY, T. M., and A. CRONQUIST. 1978. *Erechtites, in* North American Flora, Series II, Part 10, pp. 139–142.

BAYER, R. J., and G. L. STEBBINS. 1981. Chromosome numbers of North American species of *Antennaria* Gaertner (Asteraceae: Inuleae). Amer. J. Bot. **68:** 1342–1349.

———. 1982. A revised classification of *Antennaria* (Asteraceae: Inuleae) of the eastern United States. Syst. Bot. **7:** 300–313.

BEATLEY, J. C. 1963. The sunflowers (genus *Helianthus*) in Tennessee. J. Tennessee Acad. Sci. **33:** 135–154.

BELCHER, R. O. 1956. A revision of the genus *Erechtites* (Compositae), with inquiries into *Senecio* and *Arrhenechthites*. Ann. Missouri Bot. Gard. **43:** 1–85.

BIERNER, M. W. 1972. Taxonomy of *Helenium* sect. *Tetrodus* and a conspectus of North American *Helenium* (Compositae). Brittonia **24:** 331–355.

BRAUN, E. L. 1961. The Woody Plants of Ohio. Ohio State University Press, Columbus.

———. 1967. The Monocotyledoneae [of Ohio]. Ohio State University Press, Columbus.

BREHM, B. G., and M. OWNBEY. 1965. Variation in chromatographic patterns in the *Tragopogon dubius-pratensis-porrifolius* complex (Compositae). Amer. J. Bot. **52:** 811–818.

BRITTON, N. L. 1901. Manual of the Flora of the Northern States and Canada. Henry Holt, New York.

BROUILLET, L., and J. C. SEMPLE. 1981. A propos du status taxonomique de *Solidago ptarmicoides*. Canad. J. Bot. **59:** 17–21.

CANNE, J. M. 1977. A revision of the genus *Galinsoga* (Compositae: Heliantheae). Rhodora **79:** 319–389.

CLEVENGER, S., and C. B. HEISER, JR. 1963. *Helianthus laetiflorus* and *H. rigidus*—hybrids or species? Rhodora **65:** 121–133.

CLEWELL, A. F., and J. W. WOOTEN. 1971. A revision of *Ageratina* (Compositae: Eupatorieae) from eastern North America. Brittonia **23:** 123–143.

CLONTS, J. A., and S. McDANIEL. 1978. *Elephantopus, in* North American Flora, Series II, Part 10, pp. 196–202.

COLEMAN, J. R. 1971. The status of the genus *Actinomeris* Nutt (=*Verbesina* L.) as revealed by experimental hybridization. Bull. Torrey Bot. Club **98:** 327–331.

————. 1974. Experimental hybridization of *Verbesina helianthoides, V. heterophylla,* and *V. chapmanii* (Compositae). Bot. Gaz. **135:** 5–12.

————. 1977. A summary of experimental hybridization in *Verbesina* (Compositae). Rhodora **79:** 17–31.

COOPERRIDER, T. S., ed. 1982. Endangered and Threatened Plants of Ohio. Ohio Biol. Surv., Biol. Notes No. 16. The Ohio State University, Columbus.

————. 1984. Ohio's herbaria and the Ohio Flora Project. Ohio J. Sci. **84:** 189–196.

CROAT, T. B. 1970. Studies in *Solidago.* I. The *Solidago graminifolia-S. gymnospermoides* complex. Ann. Missouri Bot. Gard. **57:** 250–251.

CRONQUIST, A. 1947. Revision of the North American species of *Erigeron* north of Mexico. Brittonia **6:** 121–302.

————. 1952. Compositae, *in* H. A. Gleason, The New Britton and Brown Illustrated Flora of the Northeastern United States and Adjacent Canada, Vol. 3, pp. 323–545. The New York Botanical Garden, Bronx.

————. 1978. *Petasites, in* North American Flora, Series II, Part 10, pp. 174–179.

————. 1980. Vascular Flora of the Southeastern United States. Vol. 1. Asteraceae. Univ. of North Carolina Press, Chapel Hill.

CRUISE, J. E. 1964. Biosystematic studies of three species in the genus *Liatris.* Canad. J. Bot. **42:** 1445–1455.

CUSICK, A. W. 1981. Compass-plant (*Silphium laciniatum*) in Ohio: its historic and present-day distribution (Abstract) *in* R. L. Stuckey and K. J. Reese, eds., The Prairie Peninsula—in the "Shadow" of Transeau: Proceedings of the Sixth North American Prairie Conference, The Ohio State University, Columbus, Ohio, 12–17 August 1978. Ohio Biol. Surv., Biol. Notes No. 15, p. 262. The Ohio State University, Columbus.

CUSICK, A. W., and G. M. SILBERHORN. 1977. The vascular plants of unglaciated Ohio. Ohio Biol. Surv. Bull. N. S. **5**(4): i–x, 1–153.

DAVIDSON, R. A. 1963. Initial biometric survey of morphological variation in the *Cirsium altissimum-C. discolor* complex. Brittonia **15:** 222–241.

FAUST, W. Z. 1972. A biosystematic survey of the *Interiores* species group of the genus *Vernonia* (Compositae). Brittonia **24:** 363–378.

FERNALD, M. L. 1950. Gray's Manual of Botany. Ed. 8. American Book Co., New York.

FISHER, T. R. 1957. Taxonomy of the genus *Heliopsis* (Compositae). Ohio J. Sci. **57:** 171–191.

————. 1958. Variation in *Heliopsis helianthoides* (L.) Sweet (Compositae). Ohio J. Sci. **58:** 97–107.

————. 1959. Natural hybridization between *Silphium laciniatum* and *S. terebinthinaceum.* Brittonia **11:** 250–254.

————. 1966. The genus *Silphium* in Ohio. Ohio J. Sci. **66:** 259–263.

————. 1981. Ohio's promiscuous sunflowers. Ohio J. Sci. **81:** 146–152.

GAISER, L. O. 1946. The genus *Liatris.* Rhodora **48:** 165–183, 216–263, 273–326, 331–382, 393–412.

GLEASON, H. A. 1952. The New Britton and Brown Illustrated Flora of the Northeastern United States and Adjacent Canada. 3 vols. The New York Botanical Garden, Bronx.

GODFREY, R. K. 1952. *Pluchea,* section *Stylimnus, in* North America. J. Elisha Mitchell Sci. Soc. **68:** 238–271.

HADLEY, E. B., and D. A. LEVIN. 1967. Habitat differences in three species of *Liatris* and their derivatives in an interbreeding population. Amer. J. Bot. **54:** 550–559.

HARMS, V. L. 1974. A preliminary conspectus of *Heterotheca* section *Chrysopsis* (Compositae). Castanea **39:** 155–165.

HEISER, C. B., JR. 1954. Variation and subspeciation in the common sunflower, *Helianthus annuus*. Amer. Midl. Naturalist **51:** 287–305.

———. 1960. Notes on two ornamental sunflowers, *Helianthus multiflorus* L. and *H. laetiflorus* Pers. Baileya **8:** 146–149.

HEISER, C. B., JR., and D. M. SMITH. 1960. The origin of *Helianthus multiflorus*. Amer. J. Bot. **47:** 860–865.

———. 1964. Species crosses in *Helianthus*. II. Polyploid species. Rhodora **66:** 344–358.

HEISER, C. B., JR., D. M. SMITH, S. B. CLEVENGER, and W. C. MARTIN, JR. 1969. The North American sunflowers (*Helianthus*). Mem. Torrey Bot. Club **22:** 1–218.

JACKSON, R. C. 1956. The hybrid origin of *Helianthus doronicoides*. Rhodora **58:** 97–100.

———. 1960. A revision of the genus *Iva* L. Univ. Kansas Sci. Bull. **41:** 793–875.

JACKSON, R. C., and A. T. GUARD. 1957. Natural and artificial hybridization between *Helianthus mollis* and *H. occidentalis*. Amer. Midl. Naturalist **58:** 422–433.

JONES, A. G. 1980. A classification of the New World species of *Aster* (Asteraceae). Brittonia **32:** 230–239.

———. 1983. Nomenclatural changes in *Aster* (Asteraceae). Bull. Torrey Bot. Club **110:** 39–42.

JONES, A. G., and D. A. YOUNG. 1982. Generic concepts in *Aster* (Asteraceae): a comparison of cladistic, phenetic, and cytological approaches. Syst. Bot. **8:** 71–84.

JONES, S. B. 1972. A systematic study of the *Fasciculatae* group of *Vernonia* (Compositae). Brittonia **24:** 28–45.

JONES, S. B., and W. Z. FAUST. 1978. Vernonieae, *in* North American Flora, Series II, Part 10, pp. 180–196.

KECK. D. D. 1946. A revision of the *Artemisia vulgaris* complex in North America. Proc. Calif. Acad. Sci. 4th Ser. **25:** 421–468.

KELLERMAN. W. A. 1902. Minor plant notes, No. 4. Ohio Naturalist **2:** 179–181.

KING, R. M., and H. ROBINSON. 1970. *Eupatorium*, a composite genus of Arctotertiary distribution. Taxon **19:** 769–774.

LEVIN, D. A. 1967. An analysis of hybridization in *Liatris*. Brittonia **19:** 248–260.

LONG, R. W. 1954. Synopsis of *Helianthus giganteus* L. and related species. Rhodora **56:** 198–203.

———. 1955. Hybridization between the perennial sunflowers *Helianthus salicifolius* A. Dietr. and *H. grosseserratus* Martens. Amer. Midl. Naturalist **54:** 61–64.

———. 1958. Notes on the distribution of Ohio Compositae: I. Heliantheae, Anthemideae. Rhodora **60:** 125–128.

———. 1959. Notes on the distribution of Ohio Compositae: II. Eupatorieae, Senecioneae, Cynareae, Cichorieae. Rhodora **61:** 76–79.

———. 1961. Biosystematics of two perennial species of *Helianthus* (Compositae), II. Natural populations and taxonomy. Brittonia **13:** 129–141.

McCANCE, R. M., JR., and J. F. BURNS, eds. 1984. Ohio Endangered and Threatened Vascular Plants: Abstracts of State-listed Taxa. Ohio Dept. Natural Resources, Division of Natural Areas and Preserves, Columbus.

McGREGOR, R. L. 1968. The taxonomy of the genus *Echinacea* (Compositae). Univ. Kansas Sci. Bull. **48:** 113–142.

MEARS, J. A. 1975. The taxonomy of *Parthenium* section *Partheniastrum* DC. (Asteraceae-Ambrosiinae). Phytologia **31:** 463–482.

MILSTEAD, W. L. 1964. A revision of the North American species of *Prenanthes*. Ph.D. dissertation, Purdue University, West Lafayette, Indiana.

MONTGOMERY, J. D., and D. E. FAIRBROTHERS. 1970. A biosystematic study of the *Eupatorium rotundifolium* complex (Compositae). Brittonia **22**: 134–150.

MOORE, R. J., and C. FRANKTON. 1974. The Thistles of Canada. Biosystematics Research Institute, Canada Dept. of Agriculture, Monograph No. 10.

MORGAN, J. T. 1967. A taxonomic study of the genus *Boltonia* (Asteraceae). Ph.D. dissertation, University of North Carolina, Chapel Hill.

MORTON, G. 1978. *Tussilago*, *in* North American Flora, Series II, Part 10, p. 174.

MOSELEY, E. L. 1931. Some plants that were probably brought to northern Ohio from the west by Indians. Pap. Michigan Acad. Sci. **13**: 169–172.

OWNBEY, G. B. 1964. Natural hybridization in the genus *Cirsium*—II. *C. altissimum* x *C. discolor*. Michigan Bot. **3**: 87–97.

OWNBEY, M. 1950. Natural hybridization and amphiploidy in the genus *Tragopogon*. Amer. J. Bot. **37**: 487–499.

PIPPEN, R. W. 1978. *Cacalia*, *in* North American Flora, Series II, Part 10, pp. 151–159.

PRINGLE, J. S. 1982. The distribution of *Solidago ohioensis*. Michigan Bot. **21**: 51–57.

RICHARDS, E. L. 1968. A monograph of the genus *Ratibida*. Rhodora **70**: 348–393.

RIDDELL, J. L. 1836. A supplementary catalogue of Ohio plants. West. J. Med. Phys. Sci. **9**: 567–592.

ROBERTS, M. L. 1983. Allozyme variation in *Bidens discoidea* (Compositae). Brittonia **35**: 239–247.

———. 1985. The cytology, biology, and systematics of *Megalodonta beckii* (Compositae). Aquatic Bot. **21**: 99–110.

ROBERTS, M. L., and T. S. COOPERRIDER. 1982. Dicotyledons, *in* T. S. Cooperrider, ed., Endangered and Threatened Plants of Ohio. Ohio Biol. Surv., Biol. Notes No. 16, pp. 48–84. The Ohio State University, Columbus.

ROCK, H. F. L. 1957. A revision of the vernal species of *Helenium* (Compositae). Rhodora **59**: 101–116, 128–158, 168–178, 203–216.

ROLLINS, R. 1950. The Guayule rubber plant and its relatives. Contr. Gray Herb. **172**: 1–73.

ROWLEY, G. D. 1961. The naming of hybrids. Taxon **10**: 211–212.

———. 1964. The naming of hybrids (2). Taxon **13**: 64–65.

SEMPLE, J. C. 1977. Cytotaxonomy of *Chrysopsis* and *Heterotheca* (Compositae-Astereae): a new interpretation of phylogeny. Canad. J. Bot. **55**: 2503–2513.

———. 1978. The cytogeography of *Aster pilosus* (Compositae): Ontario and the adjacent United States. Canad. J. Bot. **56**: 1274–1279.

SEMPLE, J. C., V. C. BLOK, and P. HEIMAN. 1980. Morphological, anatomical, habit, and habitat differences among the goldenaster genera *Chrysopsis*, *Heterotheca*, and *Pityopsis* (Compositae-Astereae). Canad. J. Bot. **58**: 147–163.

SEMPLE, J. C., and L. BROUILLET. 1980. A synopsis of North American asters: the subgenera, sections and subsections of *Aster* and *Lasallea*. Amer. J. Bot. **67**: 1010–1026.

SEMPLE, J. C., and C. C. CHINNAPPA. 1980. Karyotype evolution and chromosome numbers in *Chrysopsis* (Nutt.) Ell. sensu Semple (Compositae-Astereae). Canad. J. Bot. **58**: 164–171.

SHERFF, E. E. 1955a. *Bidens*, *in* North American Flora, Series II, Part 2, pp. 70–129.

———. 1955b. *Coreopsis*, *in* North American Flora, Series II, Part 2, pp. 4–40.

———. 1955c. *Megalodonta*, *in* North American Flora, Series II, part 2, p. 147.

SHINNERS, L. H. 1947. Revision of the genus *Krigia* Schreber. Wrightia **1:** 187–206.

SIEREN, D. J. 1981. The taxonomy of the genus *Euthamia*. Rhodora **83:** 551–579.

SMITH, D. M., and A. T. GUARD. 1958. Hybridization between *Helianthus divaricatus* and *H. microcephalus*. Brittonia **10:** 137–145.

SMITH, E. B. 1976. A biosystematic survey of *Coreopsis* in eastern United States and Canada. Sida **6:** 123–215.

SMITH, E. B., and H. M. PARKER. 1971. A biosystematic study of *Coreopsis tinctoria* and *C. cardaminefolia* (Compositae). Brittonia **23:** 161–170.

SPEER, J. M 1958. The genus Aster in Ohio. M.S. thesis, The Ohio State University, Columbus.

STEYERMARK, J. A. 1951. A glabrous variety of *Silphium terebinthinaceum*. Rhodora **53:** 133–135.

————. 1963. Flora of Missouri. Iowa State University Press, Ames.

STROTHER, J. L. 1969. Systematics of *Dyssodia* Cavanilles (Compositae: Tageteae). Univ. Calif. Publ. Bot. **48:** 1–88.

STUCKEY, R. L., and J. L. FORSYTH. 1971. Distribution of naturalized *Carduus nutans* (Compositae) mapped in relation to geology in northwestern Ohio. Ohio J. Sci. **71:** 1–15.

URBATSCH, L. E. 1972. Systematic study of the *Altissimae* and *Giganteae* species groups of the genus *Vernonia* (Compositae). Brittonia **24:** 229–238.

VOSS, E. G., chm. ed. comm. 1983. International Code of Botanical Nomenclature. Bohn, Scheltema & Holkema, Utrecht/Antwerp.

VOSS, E. G., and M. W. BÖHLKE. 1978. The status of certain hawkweeds (*Hieracium* subgenus *Pilosella*) in Michigan. Michigan Bot. **17:** 35–47.

VUILLEUMIER, B. S. 1973. The genera of Lactuceae (Compositae) in the southeastern United States. J. Arnold Arbor. **54:** 42–93.

WAGNER, W. H., JR. 1958. The hybrid ragweed, *Ambrosia artemisiifolia* x *A. trifida*. Rhodora **60:** 309–316.

WATSON, E. E. 1929. Contributions to a monograph of the genus *Helianthus*. Pap. Mich. Acad. Sci. **9:** 305–475.

WEBER, W. R. 1968. Biosystematic studies in the genus *Silphium* L. (Compositae): investigations in the selected intraspecific taxa of *Silphium asteriscus* L. Ph.D. dissertation, The Ohio State University, Columbus.

WEED, C. M. 1890. The Lakeside Daisy. J. Columbus Hort. Soc. **5:** 72–73.

WEISHAUPT, C. G. 1971. Vascular plants of Ohio. A Manual for Use in Field and Laboratory. Ed. 3. Kendall/Hunt Publishing Co., Dubuque.

WELLS, J. R. 1965. A taxonomic study of *Polymnia* (Compositae). Brittonia **17:** 144–159.

# INDEX TO PLANT NAMES

The page on which a species is illustrated is in **boldface.** Forms are not indexed. Synonyms are in *italics.* Names occurring in the introduction are not indexed.